Adaptive Structures

Adaptive Structures

Engineering Applications

EDITED BY

David Wagg, Ian Bond, Paul Weaver and Michael Friswell
Faculty of Engineering, University of Bristol, UK

BICENTENNIAL
1807
WILEY
2007
BICENTENNIAL

John Wiley & Sons, Ltd

Other Wiley Editorial Offices

John Wiley & Sons Inc., 111 River Street, Hoboken, NJ 07030, USA

Jossey-Bass, 989 Market Street, San Francisco, CA 94103-1741, USA

Wiley-VCH Verlag GmbH, Boschstr. 12, D-69469 Weinheim, Germany

John Wiley & Sons Australia Ltd, 42 McDougall Street, Milton, Queensland 4064, Australia

John Wiley & Sons (Asia) Pte Ltd, 2 Clementi Loop #02-01, Jin Xing Distripark, Singapore 129809

John Wiley & Sons Canada Ltd, 6045 Freemont Blvd, Mississauga, ONT, Canada L5R 4J3

Wiley also publishes its books in a variety of electronic formats. Some content that appears in print
may not be available in electronic books.

Anniversary Logo Design: Richard J. Pacifico

British Library Cataloguing in Publication Data

A catalogue record for this book is available from the British Library

ISBN 978-0-470-05697-4 (HB)

Typeset in 10.5/13pt Times by Integra Software Services Pvt. Ltd, Pondicherry, India
Printed and bound in Great Britain by TJ International Ltd, Padstow, Cornwall
This book is printed on acid-free paper responsibly manufactured from sustainable forestry in
which at least two trees are planted for each one used for paper production.

Contents

5   Adaptive Aeroelastic Structures   137
Jonathan E. Cooper

**6   Adaptive Aerospace Structures with Smart
     Technologies – A Retrospective and Future View   163**
Christian Boller

List of Contributors

EDITORS

D.J. Wagg
Department of Mechanical Engineering
University of Bristol
Bristol
BS8 1TR
UK

I.P. Bond
Department of Aerospace Engineering
University of Bristol
Bristol
BS8 1TR
UK

P.M. Weaver
Department of Aerospace Engineering
University of Bristol
Bristol
BS8 1TR
UK

M.I. Friswell
Department of Aerospace Engineering
University of Bristol
Bristol
BS8 1TR
UK

AUTHORS

S. Aimmanee
Mechanical Engineering Department
King Mongkut's University of Technology
Thonburi Thoongkru Bangkok 10140
Thailand

B.J. Blaiszik
Theoretical and Applied Mechanics Program
Department of Mechanical Science and Engineering
University of Illinois Urbana–Champaign
Urbana
IL 61801
USA

C. Boller
Department of Mechanical Engineering
The University of Sheffield
Sheffield
S1 3JD
UK

L.F. Campanile
SMS
Laboratory 119
Swiss Federal Laboratories for Materials Testing and Research (EMPA)
Überlandstr. 129
CH-8600 Dübendorf
Switzerland

R.L. Clark
Department of Mechanical Engineering & Materials Science
Duke University
Box 90300 Hudson Hall
Durham
NC 27708-0300
USA

J.E. Cooper
School of Mechanical
Aerospace and Civil Engineering
The University of Manchester
PO Box 88
Sackville Street
Manchester
M60 1QD
UK

M.-L. Dano
Département de Genie Mécanique
Université LAVAL
Québec
Canada
G1K 7P4

S.J. Elliott
Institute of Sound & Vibration Research
University of Southampton
Highfield
Southampton
SO17 1BJ
UK

L.P. Fowler
Department of Mechanical Engineering & Materials Science
Duke University
Box 90300 Hudson Hall
Durham
NC 27708-0300
USA

B.L. Grisso
Center for Intelligent Material Systems and Structures
310 Durham Hall
Mail Code 0261
Virginia Tech
Blacksburg
VA 24061
USA

M.W. Hyer
Engineering Science and Mechanics
218 Norris Hall
Virginia Tech
Blacksburg
VA 24061
USA

D.J. Inman
Center for Intelligent Material Systems and Structures
310 Durham Hall
Mail Code 0261
Virginia Tech
Blacksburg
VA 24061
USA

T.M. Jaglinski
Institute for Shock Physics
Washington State University
PO Box 642816
Pullman
WA 99164-2816
USA

A.B. Jilani
Hewlett-Packard Company
1000 Northeast Circle Blvd.
Corvallis
OR 97330
USA

M.W. Keller
Theoretical and Applied Mechanics Program
Department of Mechanical Science and Engineering
University of Illinois Urbana–Champaign
Urbana
IL 61801
USA

R.S. Lakes
University of Wisconsin
541 Engineering Research Building
1500 Engineering Drive
Madison
WI 53706-1687
USA

S.-M. Moon
Department of Mechanical Engineering & Materials Science
Duke University
Box 90300 Hudson Hall
Durham
NC 27708-0300
USA

M.R. Schultz
Composite Technology Development, Inc.
Lafayette
CO 80026
USA

N.R. Sottos
Department of Aerospace Engineering
University of Illinois at Urbana–Champaign
216 Talbot Laboratory
MC-262
104 S. Wright Street
Urbana
IL 61801
USA

J.F.V. Vincent
Department of Mechanical Engineering
The University
Bath
BA2 7AY
UK

S.R. White
Department of Aerospace Engineering
University of Illinois at Urbana–Champaign
306 Talbot Lab
104 S. Wright Street
Urbana
IL 61801
USA

Preface

This book is based around the concept of 'adaptive structures', by which we mean engineering structures which have the ability to adapt, evolve or change their properties or behaviour in response to the environment around them. In recent years this concept has developed into a richly diverse area of research which includes topics such as structures, materials, dynamics, control, design and biological systems. The interdisciplinary fusion of these individual topic areas creates the possibility for new and exciting technological developments. These developments have been taking place in a wide range of industrial applications, but are particularly advanced in the aerospace and space technology sector.

Each chapter in this book represents the current state of the art in a particular aspect of adaptive structures, written by leading experts in their respective fields. But what about future developments beyond the current state of the art? Well, many chapters include discussions on future developments. More than this, we believe that by bringing together so many interrelated and yet diverse topics in a single volume one can get a sense of the huge future potential of this rapidly developing field of research. We hope that by viewing these combined chapters as a whole, the reader can enjoy the same sense of excitement and inspiration we felt when compiling this volume.

WHAT ARE ADAPTIVE STRUCTURES?

Humans have long been fascinated by nature's ability to build structures which adapt to their environment. In contrast, our own structures often appear inefficient, static and cumbersome. In engineering, the term 'adaptive structure' has come to mean any structure which can alter either its geometric form or material properties. These are processes which are currently much simpler than those which can be observed in nature. The terms 'smart', 'intelligent' and 'active' have all been applied to describe both materials and structures which exhibit some or all of these properties (see the selection of authored and edited texts referenced below). Increasingly, the ability to adapt to a performance demand or environmental conditions has become a key design criterion for a range of structural and mechanical systems in recent years. It is precisely this type of requirement which has become a key driver in the development of adaptive structure technology.

The adaptation process itself can be passive, active, based on material properties, control, mechanical actuation or some combination of these. As performance limits on structural systems are increasinlgy being pushed to more extreme levels, especially with respect to minimising weight, there is a strong requirement to find more efficient ways to apply adaption processes. This brings significant scientific challenges relating to structural stability, vibration, control/actuation, sensing and material behaviour.

There are many examples of adaptive structures from a broad range of engineering applications, but much of the driving force for development has come from the aerospace and space engineering sectors. The need for a high level of material performance in terms of strength, flexibility and minimal weight, coupled with the need for deployment and operation in extreme environments, has led to some of the most advanced adaptive structures currently in existence. There has also been considerable interest in new concepts such as 'morphing' wings for aircraft.

As these more advanced concepts of adaptive structures become realisable, the interaction and integration of material behaviour, control, sensing and actuation becomes ever more critical.

WHY ADAPTIVE STRUCTURES 2006?

This book forms a permanent record of the 2006 Colston Research Society Symposium on Adaptive Structures, held at the University of Bristol on 10–12 July 2006. The symposium formed part of of a wider celebration happening in Bristol during 2006 to mark the bicentenary of the birth of Isambard Kingdom Brunel (1806–1859), arguably the greatest engineer of all time. Brunel's influence on the science and application of engineering led to some of the greatest engineering achievements in history. The historic city of Bristol has special links to Brunel, with structures such as the Clifton Suspension Bridge and the SS *Great Britain* forming a prominent part of the city's engineering heritage.

Bristol retains a strong link with modern engineering as a key centre for European aerospace manufacture. Representatives from local industry took part in the symposium and a public lecture by Gordon McConnell, Senior Vice President – Engineering at Airbus UK, was given on 'Continuing the Vision – Airbus A380 and Beyond'. The focus of this symposium was to consider the direction and key challenges associated with the rapidly developing field of adaptive structures.

WHAT DID WE LEARN?

The book chapters stand alone in giving detailed information in specific topic areas. However, there are some strong 'emergent' or common themes which relate the diverse array of subject areas – from precise theoretical mechanics and control in piezoelectric devices, through advanced polymer chemistry, to the innermost workings of a Venus Fly Trap.

Firstly, it is clear that advanced material properties lie at the very heart of adaptive structures. In this book, topics covered range from chemistry to theoretical mechanics – seemingly disparate areas but crucial to the understanding of many problems. In fact this example highlights one of the key concepts to emerge from this book – integrated thinking. What do we mean by this? Any material has a chemical make-up and at the same time a mechanical behaviour. Our traditional approach to scientific research means that these two things are treated as completely separate subjects, so much so that many practitioners from either field may not even be able to communicate with each other! What is clear for adaptive structures research is that integrated understanding of a material's behaviour can lead to novel ways of exploitation.

This becomes more specific when we think about 'material' and 'structural' properties. Again, although often separated, throughout this book we see examples of 'multifunctionality' which makes this traditional separation irrelevant. In essence we need to see a blurring of the distinction between 'structure' and 'material', because when we do this new possibilities emerge which can potentially be exploited. This is often possible across the length scales. Although the dominant motivations are for aerospace and space applications (and a small amount of civil engineering) there are also many possibilities at the MEMS and nano scales.

Another common thread in this work is that of biological inspiration (or analogy/function). A clear message is that nature uses information and structure rather than energy to design its structures. Nature also makes significant use of hierarchy throughout its adaptive structures, leading quite naturally to multifunctional behaviour. It is also clear that with regard to obtaining information and acting upon it, our current sensing and control/actuation technology is some way behind that employed by nature. Again, we see that there is a strong driver towards integration of function – sensors which are also actuators, materials with integral sensors, etc. Structural health monitoring is relatively new in engineering, but an entirely natural (and essential) process for biological systems. Ways of efficiently closing the control loop to provide feedback continues to challenge our traditional engineering approach.

Imparting some degree of self-healing to an engineering structure/material is perhaps a prime example of how research is attempting to bridge the divide between the synthetic and organic. One can envisage such a functionality offering real benefit across a wide range of engineering applications; however, replicating the subtleties of the natural world continues to pose significant challenges.

Overall, there is a strong sense that we need to challenge existing ways of thinking: concepts, assumptions and design approaches. Throughout this book the reader will see examples of this type of new and questioning approach. But these thought processes are not just idle speculation – almost every one is backed up with high-quality experimental validation.

Arguably, had Brunel been alive today he would have been a champion of the thinking that is encapsulated within the work presented here. Never one to shy away from applying the latest technologies available, or indeed finding his own solutions to problems, Brunel would no doubt approve of the theme, topics and findings presented herein.

ACKNOWLEDGEMENTS

We are indebted to a wide range of people who gave their time and effort to make both the symposium and this book a success. In no preferential order we should thank Victoria Child, Charlotte Eve, Sarah Hugo, Nick Lieven, Gordon McConnell, Nihal Malik, Julie Etches, Richard Trask, Cezar Diaconu, Andy Kinsman, Jon Scholey, John Bracey, Joan Lovell and all the staff at Burwalls Conference Centre. Special thanks should go to David

Palmer, Sarah Powell, Wendy Hunter and Kelly Board at John Wiley & Sons for their help and support in publishing this book. Finally, we would like to thank the Colston Research Society and the UK Engineering and Physical Sciences Research Council for providing financial support for this symposium.

David Wagg
Ian Bond
Paul Weaver
Michael Friswell
University of Bristol, August 2006

BIBLIOGRAPHY

Addington, M. & Schodek, D.L. (2004), Smart Materials and Technologies in Architecture, Architectural Press.

Adeli, H. & Saleh, A. (1999), Control, Optimization, and Smart Structures: High-Performance Bridges and Buildings of the Future, John Wiley & Sons, Ltd.

Carman, G. & Garcia, E., Eds. (1993), Adaptive Structures and Material Systems, American Society of Mechanical Engineers.

Clark, R.L.; Saunders, W.R. & Gibbs, G.P. (1998), Adaptive structures: dynamics and control, John Wiley & Sons, Ltd.

Ewins, D. & Inman, D., Eds. (2003), Structural Dynamics @ 2000: Current Status and Future Directions, Research Studies Press.

Gawronski, W.K. (2004), Advanced Structural Dynamics and Active Control of Structures, Springer.

Guran, A. & Inman, D., Eds. (1995), Wave Motion, Intelligent Structures and Nonlinear Mechanics, World Scientific.

Guran, A. & Inman, D.J., Eds. (1995), Smart Structure, Nonlinear Dynamics and Control, Prentice Hall.

Inman, D. (2006), Vibration with Control, John Wiley & Sons, Ltd.

Preumont, A. (2002), Vibration Control of Active Structures: An Introduction, Kluwer Academic.

Redmond, J. & Main, J., Eds. (2000), Adaptive Structures and Material Systems, American Society of Mechanical Engineers.

Singh, J. (2005), Smart Electronic Materials: Fundamentals and Applications, Cambridge University Press.

Srinivasan, A.V. (2000), Smart Structures: Analysis and Design, Cambridge University Press.

Utku, S. (1998), Theory of Adaptive Structures: Incorporating Intelligence into Engineered Products, CRC Press.

Wada, B.K.; Fanson, J.L. & Miura, K., Eds. (1991), First Joint US/Japan Conference on Adaptive Structures: November 13–15, 1990 Maui, Hawaii, USA, CRC Press Inc.

1
Adaptive Structures for Structural Health Monitoring

Daniel J. Inman and Benjamin L. Grisso

Center for Intelligent Material Systems and Structures, Department of Mechanical Engineering, 310 Durham Hall, Mail Code 0261, Virginia Tech, Blacksburg, VA24061, USA

1.1 INTRODUCTION

For some time the adaptive structures community has focused on trans-ducer effects, and the closest advance into actually having a structural system show signs of intelligence is to include adaptive control imple-mented with a smart material. Here we examine taking this a step further by combining embedded computing with a smart structural system in an attempt to form an autonomous sensor system. The focus here is based on an integrated structural health monitoring system that consists of a completely wireless, active sensor with embedded electronics, power and computing. Structural health monitoring is receiving increased attention in industrial sectors and in government regulatory agencies as a method of reducing maintenance costs and preventing disasters. Here we propose and discuss an integrated autonomous sensor 'patch' that contains the following key elements: sensing, energy harvesting from ambient vibration and temperature, energy storage, local computing/decision making, memory, actuation and

Adaptive Structures: Engineering Applications Edited by D. Wagg, I. Bond, P. Weaver and M. Friswell
© 2007 John Wiley & Sons, Ltd

wireless transmission. These elements should be autonomous, self-contained and unobtrusive compared to the system being monitored. Each of these elements is discussed as a part of an integrated system to be used in structural health monitoring applications.

In addition, the concept of using smart materials in a combined monitoring and self-healing function is briefly discussed. This chapter concludes with some thoughts on the way forward in monitoring which is a subset of the newly formed area called 'autonomic structures' and includes a short introduction to such systems.

Autonomous sensing requires the integration of a number of subsystems: power, sensor material, actuation material, energy management, telemetry and computing. This chapter discusses one such solution to building an autonomous sensing system as well as steps taken to further integrate such a system into a load-bearing adaptive structure. The basic idea of the autonomous sensing system proposed here is summarized in Figure 1.1.

The proposed sensing system must have the following components in order to function autonomously. First, it must be built around a transducer material that performs the basic sensing function. For the example discussed here, this material consists of a piezoceramic, which produces an electric field when strained (see, for instance, Dosch *et al.*, 1994). The electric field is then converted to a voltage, which is proportional to local strain and can be used to measure local displacement or velocity. Here, however, we are interested in measuring the electrical impedance of the sensing piezoceramic (PZT in this case) as discussed below. Figure 1.1 also indicates that the PZT serves as an actuator as well. Actuation is needed because many of

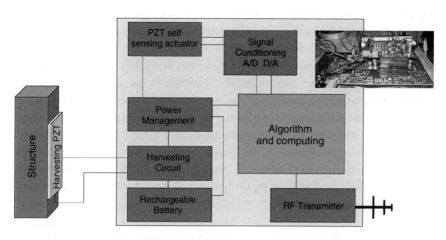

Figure 1.1   A proposed autonomous sensing system

the best algorithms for structural health monitoring (SHM) require a known input (Doebling *et al.*, 1998) in order to form measurements resembling a transfer function. At the very least, input–output measurements contain much more information than output only measurements. The existence of this actuation element separates the proposed active sensing system from many of the wireless sensing systems proposed by others (such as the Mote system). The circuit of Dosch *et al.* (1994) allows 'self-sensing actuation' and results in a reduction in the number of required components, reducing the size and weight requirements.

The second key element in the proposed autonomous sensing scheme is the use of a local computing platform. In a review of smart sensing technology for civil applications, smart sensors are defined as sensors which contain an onboard microprocessor giving the system intelligence capabilities (Spencer *et al.*, 2004). Several sensor platforms have incorporated microprocessors for the purpose of power management and signal conditioning using off-the-shelf chips. The system here takes the approach that (a) it takes less energy to compute than to transmit raw data, and (b) at some point during the sensor's life, it may be desirable to remotely change the algorithm used to determine damage. This is also the area in which further autonomy can be gained by enabling the sensor to make decisions. The philosophy of this approach is to make all the calculations at the sensor location and to broadcast only a limited amount of information in the form of a decision. Localized computing and decision again separates the proposed system from many of the previous efforts in the literature (Straser and Kiremidjian, 1998; Giurgiutiu and Zagrai, 2002; Lynch *et al.*, 2002, 2003). However, Lynch *et al.* (2004a, b) also use at-the-sensor computing to perform a time series analysis and broadcasts results, rather than raw data streams. A Berkeley–Mote platform is also used as a basis for a wireless structural health monitoring system with an embedded damage detection algorithm (Tanner *et al.*, 2003). A main difference between the approach presented here and other approaches is that they use a standard operating system whereas the goal here is to diminish the operating system to further reduce the power required to run the system.

The third key element of the system of Figure 1.1 is the power harvesting, management and storage system. Most systems to date use batteries as the source, and our goal here is to extend the autonomy of the sensor system by using various energy harvesting methods, power management and energy storage devices. The transmission device is taken as a standard off-the-shelf system here (see Lynch and Koh, 2005), and no new results are offered in the telemetry area. The main thrust of the work proposed here is to examine energy conservation through using a digital signal processor (DSP) platform without using an operating system (which tends to waste energy).

1.2 STRUCTURAL HEALTH MONITORING

Damage prognosis (DP) is the prediction in near real time of the remaining useful life of an engineered system given the measurement and assessment of its current damaged (or aged) state and accompanying predicted performance in anticipated future loading environments (Inman *et al.*, 2005). Self-healing can be thought of as structural repair of damage. A key element in damage prognosis and self-healing is obviously that of structural health monitoring (SHM). The added effort in damage prognosis is the concept of organizing the ability to make a decision based on the current assessment of damage by assuming future loads and predicting how the damaged system will behave. This prediction is then used to make a decision about how to use the damaged structure (or if to use it) going forward. A military aircraft hit by enemy fire gives a simple example of a prognosis system. The ideal prognosis system would detect the damage and inform the pilot if he/she should bail out, ignore the damage or perhaps continue to fly by under reduced flight performance. The battery indicator on a laptop performs a similar prediction in the sense that it measures current usage and estimates the remaining time left before required shutdown.

The added effort in self-healing is repairing the damage to return the structure to a usable state. A simple example is given below of a self-healing mechanism, while 'Self-healing composite materials' in dealt with in Chapter 9 of this volume. In the example given below of a self-healing bolted joint, there is a need to know the extent of the damage before self-repair can begin. Again, the concept of determining the state of the structure's health and the extent of its damage is a key element in the process. In this sense, damage prognosis and damage mitigation are natural extensions to SHM and can be viewed as the next steps.

In order of increasing difficulty, damage monitoring and prognosis problems can be categorized in the following stages of increasing difficulty:

1. Determining the existence of damage.
2. Determining the existence and location of damage.
3. Determining the existence, location and characterization (quantification) of damage.
4. All of the above and predicting the future behavior under various loads (damage prognosis).
5. All of the above and mitigating the effects of damage (self-healing structures).

6. Combining problems 1, 2, 3 or 4 with smart materials to form self-diagnosing Structures.

7. Combining the above with adaptive structures to form autonomous, self-repairing structures (autonomic structures).

Adaptive materials, or smart materials, and structures integrate very nicely into all seven of these problems. In the following, several examples are given to illustrate the effect that integrating these two disciplines has on solving problems arising in damage prognosis and mitigation, with the goal of eventually producing an entirely standalone chip fully integrated into a structure.

There are numerous SHM algorithms. A review of the SHM literature (Doebling *et al.*, 1998; Sohn *et al.*, 2003; Inman *et al.*, 2005) indicates that the main drawbacks and issues of the current SHM methods include:

1. *Spatial aliasing*: Conventional monitoring is accomplished with a limited number of sensors dispersed over a relatively large area of a structure providing poor spatial resolution and thus is only capable of detecting fairly significant damage.

2. *Cabling issues*: As a new generation of sensing technologies and sensor arrays pushes the limits of scale, the cabling and bookkeeping of sensor arrays has become an issue. Although wireless communication technology can provide a partial solution to this problem, *unwavering power supply to the transmitter remains largely unsolved*.

3. *Environmental issues*: Varying environmental and operational conditions produce changes in the system's dynamic response that can be easily mistaken for damage.

4. *Integration issues*: The predominant approach is to design separate systems leading to inefficiencies and reduced capabilities that could be increased through an integrated design philosophy.

Much activity has emerged in the area of wireless sensing (see, for instance, Lynch *et al.*, 2003). However, few have focused on the power requirements or on the integration of the algorithms into the choice of sensing hardware. In summary, the basic roadblock in adapting SHM methods in practice is that commercial sensing systems have not been developed with the intent of specifically addressing these drawbacks. The need to develop a system that goes beyond the laboratory demonstration and can be deployed in the field on real-world structures necessitates the goal of this effort: *that new sensing hardware must be developed in conjunction with software interrogation algorithms.*

The SHM algorithm used here is called the impedance method, was introduced by Liang *et al.* (1994), was used extensively over the last 10 years and is described next. Other algorithms, such as Lamb wave methods or vibration-based methods, can also be used, but, for the sake of simplicity and example, only impedance methods are discussed here.

1.3 IMPEDANCE-BASED HEALTH MONITORING

Impedance-based health monitoring techniques utilize small piezoceramic (PZT) patches attached to a structure as self-sensing actuators to simultaneously excite the structure with high-frequency excitations and monitor changes in the patch electrical impedance signature (Park *et al.*, 2003). Since the PZT is bonded directly to the structure of interest, it has been shown that the mechanical impedance of the structure is directly correlated with the electrical impedance of the PZT (Liang *et al.*, 1994). Thus, by observing the electrical impedance of the PZT, assessments can be made about the integrity of the mechanical structure.

The impedance-based health monitoring method is made possible through the use of piezoelectric patches bonded to the structure that act as both sensors and actuators on the system. When a piezoelectric is stressed, it produces an electric charge. Conversely, when an electric field is applied, the piezoelectric produces a mechanical strain. The patch is driven by a sinusoidal voltage sweep. Since the patch is bonded to the structure, the structure is deformed along with it and produces a local dynamic response to the vibration. The area one patch can excite depends on the structure and material. The response of the system is transferred back from the piezoelectric patch as an electrical response. The electrical response is then analyzed and, since the presence of damage causes the response of the system to change, damage is shown as a phase shift and/or magnitude change in the impedance.

The solution to the wave equation gives the following equation for electrical admittance as a function of the excitation frequency ω:

$$Y(\omega) = i\omega a \left(\bar{\varepsilon}_{33}^T (1 - i\delta) - \frac{Z_s(\omega)}{Z_s(\omega) + Z_a(\omega)} d_{3x}^2 \hat{Y}_{xx}^E \right) \qquad (1.1)$$

In Equation (1.1), Y is the electrical admittance (inverse of impedance), Z_a and Z_s are the PZT material's and the structure's mechanical impedances, respectively, Y_{xx}^T is the complex Young's modulus of the PZT with zero electric field, d_{3x} is the piezoelectric coupling constant in the arbitrary x

direction at zero stress, ε_{33}^T is the dielectric constant at zero stress, δ is the dielectric loss tangent of the PZT, and a is a geometric constant of the PZT. This equation indicates that the electrical impedance of the PZT bonded onto the structure is directly related to the mechanical impedance of a host structure.

The impedance method has many advantages compared to global vibration-based and other damage detection methods. Low excitation forces, combined with high frequencies (typically greater than 30 kHz), produce power requirements in the range of microwatts. The small wavelengths at high frequencies also allow the impedance method to detect minor local changes in structural integrity and, in some cases, imminent damage.

The impedance method has been used successfully to warn of impending damage in a number of different experiments and field tests. These range from simple laboratory tests to illustrating the method on the NASA Space Shuttle launch tower.

Traditionally, the impedance method requires the use of an impedance analyzer. Such analyzers are bulky and expensive, and are not suited for permanent placement on a structure. With the current trend of SHM heading towards unobtrusive self-contained sensors, the first steps in meeting the low-power requirements resulted in the MEMS-Augmented Structural Sensor (MASSpatch) (Grisso et al., 2005).

The use of a relatively small resistive circuit instead of the impedance analyzer was made possible by Peairs et al. (2004a). The idea of the 'low-cost' impedance-measuring device is to remove the need for a bulky analyzer and replace it with an operational-amplifier-based device. Impedance measurements can then be generated utilizing an FFT analyzer and small current measuring circuit. FFT analyzers, such as those used in modal analysis, are much more common and less expensive than impedance analyzers and are available on a chip.

In fact, Analog Devices has also recently introduced impedance measurement devices in chip format. The AD5933 has a 1 MSPS sampling rate and also comes in an evaluation board format. A prototype similar to MASSpatch has been developed using the AD5933 evaluation board, an ATmega128L microprocessor, and Xbee radios for wireless communications (Mascarenas et al., 2006). Using the microprocessor to control the evaluation board, bolt loosening was detected in a frame structure.

In contrast, the system described here is based on a single board computer system, which interrogates a structure utilizing a self-sensing actuator and the low-cost impedance method. All the structural interrogation and data analysis are performed in near real time at the sensor location. Wireless transmissions alert the end user to any harmful changes in the structure. The first version of this (MASSpatch) had some limitations. The algorithm, written in C, to perform the impedance method was utilized as an executable in the DOS operating system. When using an operating system, much of the processing power is used to run the actual system, as well as the algorithm. Determining how much energy is used for calculating the actual algorithm is difficult. Also, a digital-to-analog converter (DAC) was never fully incorporated into the system and reliance on an external function generator was needed for structural excitation. For these reasons, a new processing device must be used in order to optimize the prototype. The current system is based on a digital signal processor (DSP) platform. The benefits of this new system proposed here are discussed, along with current research and the path forward to a complete standalone SHM system.

1.4 LOCAL COMPUTING

To implement the impedance-based SHM method in a field-deployable setup, hardware is assembled as shown in Figure 1.2. Using the low-cost technique, accurate approximations of the structural impedance can be determined without complex and expensive external electronic analyzers. As shown in Figures 1.2 and 1.3, all of the hardware needed to utilize the impedance method is condensed into a single stacked board configuration. A description of each of the components follows.

This prototype is based on a TMS320C6713 DSK evaluation DSP module from Texas Instruments (Texas Instruments, 2005b). The DSP has an internal system clock speed of 225 MHz, 192 kB of internal memory, and external synchronous dynamic random access memory (SDRAM) of 16 MB. With a large amount of external memory, the memory space is partitioned into two major sections: samples for DAC output, and samples from the analog-to-digital converter (ADC). As shown in Figure 1.2, the ADC, DAC and

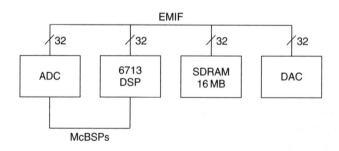

Figure 1.2 A diagram of the proposed hardware configuration

SDRAM all share an external memory interface (EMIF). The DSP controls the ADC by means of a multichannel buffered serial port (McBSP) acting in general purpose input–output (GPIO) mode.

Two more evaluation boards from Texas Instruments are used as the ADC and DAC. The ADS8364 EVM ADC board has six channels of input and a 250 kHz sampling rate (Texas Instruments, 2002). Conversion resolution for the ADC board is 16 bits. For the DAC, a TLV5619-5639 EVM board is used with a 5639 DAC (Texas Instruments, 2001). The DAC evaluation board has two outputs and a maximum sampling rate of 1 MHz at 12 bit resolution. The physical orientation of the DSP kit, ADC and DAC can be seen in Figure 1.3 (see also Table 1.1).

The wireless transmitter and receiver are used to indicate the current state of damage for the structure of interest. The transmitter sends a quantified amount of damage, and the receiver displays this value on a host computer. The current prototype uses Radiometrix RX2M-458-5 and TX2M-458-5 wireless sensors as the receiver and transmitter (Radiometrix Ltd., 2005).

The operational flow of the current prototype allows SHM to be performed all with one piece of hardware. The DSP board controls the entire operation. An excitation signal is sent from the DAC board simultaneously to the ADC board and the structure of interest. The ADC reads the voltage signal from the DAC and simultaneously reads the voltage across the sensing resistor

Figure 1.3 The prototype is shown with the DSP on the bottom followed by the DAC and ADC

Table 1.1   Specifications for the prototype of Figure 1.3

Processing/programming		Sensing/sampling		Wireless transmission	
Processor	TMS320C6713 225 MHz floating point DSP	ADC resolution	16 bit	Usable range	Over 1 km
Internal memory	192 KB	Max sampling frequency	250 kHz per channel 750 kHz paired	Operating frequencies	433.05– 434.79 MHz
External memory	16 MB SDRAM	Sensor types and ranges	6 analog at up to +10 to −10 V	Data bit rate	5 kbps

Actuation		Dimensions	
DAC resolution	12 bit	DSP board	22 × 11.5 cm
Max sampling frequency	1 MHz	DAC board	13.5 × 8.5 cm
Channels	2 up to 4 V	ADC board	10 × 8.5 cm
		Height	4.5 cm

(seen in the foreground of Figure 1.3) of the low-cost impedance circuit. After 10 excitation cycles, the signals are averaged, a FFT is performed, and one impedance measurement is generated. The first two measurements generated are baseline impedance curves. Once the baseline is stored, each measurement is compared to the baseline to determine by means of a damage metric whether there is damage in the structure.

Impedance signatures are, in general terms, simply frequency response functions (FRFs). They have the general appearance of FRFs, as seen in Figure 1.6 below. By monitoring the changes in the peaks of these FRFs, a simple damage algorithm can be used to quantify the amount of change in the peaks and thus the amount of damage in the structure. In this case, a variation of the root mean square deviation is used as the damage metric (Park *et al.*, 2003).

In order to excite the structure of interest, a sine cardinal, or simply sinc, was used as the DAC output. The sinc function has the unique property in that its Fourier transform is a box. Having a uniform value in the frequency domain allows for a band of frequency content in one pulse. The sinc function is based on a fundamental frequency and then frequencies which build upon the fundamental, as shown in Figure 1.4. By slightly altering the fundamental

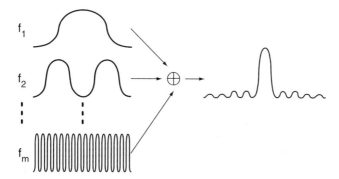

Figure 1.4 A diagram showing the sinc function (on the right) and how the function is built

frequency each time a pulse is sent out, the averaged spectrum is even smoother. The sinc function is described as (Schilling and Harris 2005)

$$\text{sinc} = \frac{\sin(x)}{x} \tag{1.2}$$

By using a sinc function instead of exciting the structure with discrete frequencies, more frequencies can be excited in the same amount of time. The auto spectrum of the output signal will also be a straight line over all the frequencies excited. Other advantages of the sinc function include needing less memory space and less traffic in the external interface, as well as lower power consumption in the DSP, ADC and DAC, a key factor for self-powered systems.

1.5 POWER ANALYSIS

Currently, the prototype runs off of DC power supplies. In a permanent setting, the prototype will operate off of battery power recharged by harvested ambient energy. To optimize the battery life and minimize the required maintenance schedule, both piezoelectric and thermal-based power harvesting can be utilized to recharge batteries (see below). Piezoelectric materials have the unique property of being able to transform mechanical strain into an electric charge. By using this property, piezoelectrics can harvest energy by using a system's own ambient motion, transform this mechanical kinetic energy into electrical potential, and store the electrical energy, power devices, or recharge a battery using power harvesting circuitry (Sodano *et al.*, 2003).

In order to prepare the prototype to be completely run off of a battery and power harvesting, a complete power analysis is done of the current system. In the prototype, the DSP board supplies power to the DAC board, which in turn supplies power to the wireless transmitter. Due to the connectivity of these systems, the power consumption of the DAC and transmitter cannot be exactly determined. However, they can be estimated. According to specifications, the maximum current the transmitter consumes is 100 mA at 5 V_{DC}, or 0.05 W (Radiometrix Ltd., 2005). The transmitter is supplied with 3.3 V_{DC}, and only sends out a very small signal, so 0.33 W (100 mA at 3.3 V_{DC}) is a high estimate. The DAC board is stated to consume 170 mA at 3.3 V_{DC} and 150 mA at 5 V_{DC}, or a range of 0.561 to 0.825 W (Texas Instruments, 2001). In this setup, the DAC board is being operated at 5 V_{DC}.

The DSP board is supplied with a 5 V_{DC} power supply, so the DSP, DAC and wireless transmitter can be measured as a group. When the system is fully turned on, but the algorithm is not being performed, the DSP requires 470 mA at 5 V_{DC}, which is 2.35 W. While the whole impedance-based SHM operation is being performed, including wireless transmission, the current draw increases to 570 mA, giving a power of 2.85 W. So, wireless transmissions are shown not to be a significant drain on the power supply. All of these measurements are instantaneous power, but the current draw remained almost constant during a complete operational cycle.

The ADC has its own ±12 V_{DC} power supply. During operation, the ADC requires 60 mA, yielding 1.44 W of power. So, the total amount of power required for the prototype to completely perform impedance-based SHM is 4.29 W. A summary of the power analysis can be seen in Table 1.2. Comparatively, the MASSpatch prototype used around 4.5 W of power (Grisso *et al.*, 2005); 4.29 W does not seem like a significant reduction considering the advances in hardware and excitation efficiency, but the previous (MASSpatch) prototype relied on an external function generator to provide excitations. MASSpatch did not include its own DAC, and the function generator used was plugged into a wall outlet and consumes a considerable amount of power.

Table 1.2 The power consumption for the prototype components

Power analysis	
Wireless (estimate)	0.33 W (max)
DAC (estimate)	0.825 W (max)
DSP, DAC, wireless group	2.85 W during operation
ADC	1.44 W
TOTAL	4.29 W

Even with 4.29 W of power, the prototype is capable of being run solely off of battery power and piezoelectric power harvesting. For instance, if the prototype was being continuously run for 10 minutes or more (a rather excessive time), 10 1.2 V, 200 mAh capacity batteries could supply more than enough energy to the system. A 1 Ah capacity means that a battery will last for 1 h if it is subjected to a discharge current of 1 A. A 200 mAh battery can be recharged to 90 % capacity in 1.2 h with a random vibration signal at 0 to 500 Hz if a 6.35 × 2.375 inch PZT is used (Sodano *et al.*, 2003).

The prospect for success of the proposed autonomous sensor can be captured in some simple energy accounting. We require

$$E_H(t) + E_H(0) > E_C(t)$$

in which $E_H(t)$ is the total energy harvested until time t from the beginning of a sensor's cycle, $E_H(0)$ is the initial energy stored in storage elements at the beginning of the cycle, and $E_C(t)$ is the total energy consumed during the period for data collection, computation, transmission. If the above equation cannot be met at a certain time instant t, then the duty cycle, environment and monitoring task will not work with the proposed autonomous sensor.

1.6   EXPERIMENTAL VALIDATION

To validate our prototype concept, the system's capabilities are demonstrated in the laboratory. A bolted joint, as seen in Figure 1.5, is tested for the initial experiments. The bolted joint structure consists of two aluminum beams connected with four bolts. A piezoelectric patch is attached to this structure; the piezoelectric acts as a self-sensing actuator. Damage is induced in the bolted joint by tightening or loosing one or more of the bolts.

Using traditional impedance techniques (a HP 4194A impedance analyzer), a standard for the bolted joint experiment is generated for comparison to results from the prototype. Initial bolted joint testing shows that the impedance method readily detects damage induced by loose bolts. Slightly

Figure 1.5   Images of the bolted joint and piezoelectric patch are shown

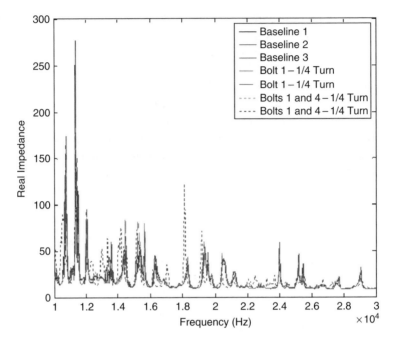

Figure 1.6 The baseline of the real value of impedance vs. frequency and damaged impedance signatures for the bolted joint are shown

loosening only one of four bolts significantly changes the impedance signature. Figure 1.6 shows the impedance curves generated using an impedance analyzer.

As displayed in Figure 1.6, the peaks of the impedance signature change as damage is introduced to the structure by loosening bolts. The more the structure is damaged, the more the peaks shift from the baseline. A frequency range of 10–30 kHz was selected for easy comparison to the prototype. A variation root mean square deviation (RMSD) damage metric is utilized to analyze changes in these peaks and determine the amount of damage present. Figure 1.7 displays this damage metric in bar chart form.

In Figure 1.7, the first two bars compare the second and third baselines (healthy measurements) to the first baseline. The next two groups of bars compare the next two damage cases, the loosening of bolt one and the combination of bolts 1 and 4.

Using the same bolted joint, the prototype could be directly compared to standard impedance measurement methods. Code Composer Studio software allows for visualization of what the damage detection algorithm is doing in the DSP core (Texas Instruments, 2005a). At each step in the algorithm, the real

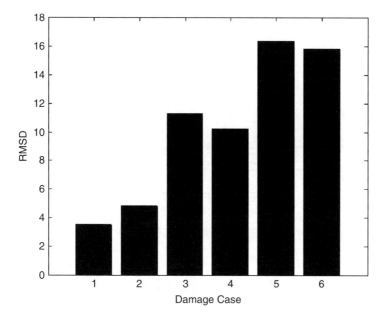

Figure 1.7 The damage metric compares the baselines and damaged curves

impedance measurement (as data is acquired) is displayed along with the baseline, the averaged real impedance measurement used to compare to the baseline, the original DAC sinc function output, and the ADC sampled output. The most important part of the display is the damage metric value, which is updated with each measurement to indicate how much damage is present in the structure. Impedance measurements are taken over the range of 10–30 kHz.

The spectrum of the output can also be displayed, as shown in Figure 1.8. As expected from a sinc function, the auto spectrum is a flat line indicating that every frequency of interest is being excited. In Figure 1.8, it should be noted that 512 frequency components are displayed, representing 0 to 64 200 Hz. In reality, only half of the spectrum is used, so 256 frequency lines represent 0 to 32 100 Hz.

Initially, measurements are taken with all of the bolts completely tightened. With no damage to the structure, the baseline and damaged impedance signature should be the same. As Figure 1.9 shows, the impedance curves for the new measurement and original baseline are almost identical. Figure 1.9 is generated by Code Composer Studio, and allows for graphical displays of what is actually occurring at specific memory locations in the hardware. All of the computations are performed on the DSP, and the graphs just show the results. The damage metric value displayed is 0.02.

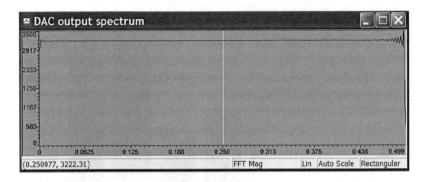

Figure 1.8   The auto spectrum is displayed for the sinc function

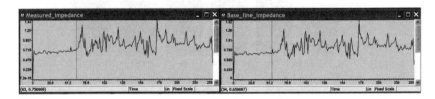

Figure 1.9   The measurement with no damage to the structure is compared to the baseline

One interesting thing to note is that the impedance signatures from Figure 1.9 and Figure 1.6 are very similar. Both show a good number of peaks in similar locations over the range of 10–30 kHz. Also, the frequency is displayed by a frequency index, i, from 0 to 256, where $i = 256$ corresponds to 32 100 Hz. The RMSD is then taken over the frequency indices of $i = 79$ to 256 (or 9906–32 100 Hz), even though the whole frequency range from 0 to 32 100 Hz is displayed. Now, a small amount of damage was induced on the bolted joint by loosening one of the bolts a quarter turn. With just this small amount of damage, the prototype easily recognizes the difference as shown in the peak changes of the measured impedance seen in Figure 1.10.

Comparing the two curves in Figure 1.11, the damage metric increased to 0.13. This is a 550 % change from the original damage metric. With only a little bit of damage, the damage metric easily indicates that the structure has changed. Next, a second bolt was also slightly loosened. Figure 1.11 displays the difference for this damage case.

As seen in Figure 1.11, with even more damage, the peaks of the measured impedance signature change even more. The damage metric also notices the change and calculates a new value of 0.21. Utilizing a bolted joint, the

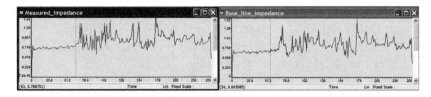

Figure 1.10   The impedance signature for slightly loosening one bolt is compared to the baseline

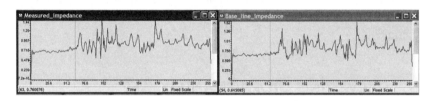

Figure 1.11   The impedance signature for slightly loosening two bolts is compared to the baseline

prototype has successfully detected varying amounts of damage. It is even more promising is that the results are very comparable to an analysis performed with standard impedance measuring equipment, a HP 4194A impedance analyzer.

As another method of validating the prototype device, frequencies displayed by both the HP 4194A impedance analyzer (Figure 1.6) and the new prototype (Figure 1.9) are directly compared. Using the impedance analyzer, data was taken in the range of 10 000–32 100 Hz. Table 1.3 shows a comparison between selected peaks shown in both Figures 1.6 and 1.9.

Table 1.3   A frequency-by-frequency comparison of the prototype

HP 4194A (Hz)	Prototype (Hz)	Difference (Hz)	Difference (%)
10 663	10 532.9	130.1	1.22
13 094	13 040.8	53.2	0.41
15 580	15 548.6	31.4	0.20
18 011.25	17 931	80.25	0.45
21 215.75	21 191.2	24.55	0.12
21 547.25	21 442	105.25	0.49
26 575	26 457.7	117.3	0.44
28 840.25	28 714.7	125.55	0.44
31 897	31 849.5	47.5	0.15

Obviously, this is only a small sampling of the frequency peaks between 10 and 32 kHz. Also, as is expected, each device is slightly more sensitive to some peaks than the other, so some peaks may be missed simply as a function of the frequency resolution. However, even when some peaks may appear to be well apart from one another, they are generally well within the frequency resolution of the machines being used. The impedance analyzer takes data from 10 to 32.1 kHz in 400 points, yielding a frequency resolution of 55.25 Hz. The prototype has a frequency resolution of 125.39 Hz. The percent difference is taken with respect to the HP 4194A impedance analyzer, which is assumed to be the true value for these experiments.

1.7   HARVESTING, STORAGE AND POWER MANAGEMENT

In order for autonomous and wireless operation to function, adequate power must be available. Batteries provide sufficient energy to run most wireless transmission systems and computational devices. However, replacing batteries greatly reduces the level of autonomy and limits sensor placement. Hence we propose to develop a system that will recharge itself using ambient vibration and thermal gradients. Some preliminary harvesting results are presented that illustrate the feasibility of recharging batteries from ambient energy.

There are many previous efforts in energy harvesting, especially using the piezoelectric effect. A summary is presented by Sondano et al. (2004b). The book by Roundy et al. (2004) provides a nice introduction to energy harvesting using vibration energy at resonance. The approach by Write et al. is to use resonance as opposed to random excitation used by Sodano et al. Other notable efforts are by Roundy (2005), who again focuses on using resonance to magnify the amount of captured energy, and Guyomar et al. (2005), who use nonlinear circuits to enhance the amount of energy captured. Here we take the approach that random vibration energy is available and that using resonance is not practical.

Depending on the thermal gradients in a given application, harvesting energy from surrounding temperature gradients can provide an order of magnitude more energy than vibration-based harvesting. Thermal energy harvesting represents an important and potentially large source of energy. When propulsion and electronics are operating, heat is generated. Some of this 'waste' energy can be harvested using thermoelectric means, simultaneously keeping these components cool. Also, daily heating and cooling provides substantial temperature gradients, which can be exploited. The power generated by thermal gradients can be approximated by

$$P_{thermal} \approx \eta A \left(T_{source} - T_{reject} \right)$$

In addition, thermal gradients exist near buildings and machinery (Sodano *et al.*, 2004a,b). Until recently, most solid-state methods relied on large temperature differences to increase the efficiency of the conversion process. Low-energy phenomena are abundant in many waste heat environments around machines and buildings. Thermoelectric coolers (TECs) are designed and manufactured to be most efficient at low temperatures (200 °C), so these devices are excellent thermoelectric generators for low-energy sources. Initial experiments show that power harvested from the ambient vibration and thermal gradients surrounding an internal combustion engine (such as a generator in a village) vary from 18 (vibration) to 70 mW (thermal).

1.7.1 Thermal Electric Harvesting

First consider the problem of capturing ambient waste heat that would be available from aircraft engines, boilers or furnaces. Thermoelectric generators (TEGs) use the Seebeck effect, which describes the current generated when the junction of two dissimilar metals/semiconductors experiences a temperature difference. Using this idea, numerous p-type and n-type junctions are arranged electrically in series, and thermally in parallel, to construct the TEG. Thus, if a thermal gradient is applied to the device, it will generate an electric current that can be utilized to power other electronics. By implementing power harvesting devices, autonomous portable systems can be developed that do not depend on traditional methods for providing power, such as the battery with its limited operating life. The idea to use thermoelectric devices to capture ambient energy from a system is not new. However, TEGs have typically been used simply to determine the extent of power capable of being generated rather than investigating applications and uses of the energy. Furthermore, the majority of previous research efforts have utilized liquid heat exchangers or forced convection to significantly improve heat flow and power generation, but require complex cooling loops and systems. These previous studies commonly do not consider the amount of energy applied to the cooling system and therefore only report gross levels of power. In the present study, thermoelectric generators will be used as power harvesting devices that do not have an active heat exchanger but function as a completely passive energy scavenging system. The motivation for investigating a passive power generation device stems from the need to identify effective power sources for the development of self-powered wireless SHM systems. These systems could be placed in a desired location without regular replacement of batteries or maintenance as most wireless devices currently require.

To simulate the energy available in these locations, a hot plate was used. A heat sink was attached to the hot plate to allow more energy to be removed from the cold side of the TEG and facilitate a larger power output. The purpose of this study is to investigate completely passive power harvesting methods, so the systems did not include a means of providing forced convection. The energy produced by the TEG would be greatly increased by forced convection, but active cooling would require additional energy.

The second source of ambient thermal energy tested was waste heat from engines, boilers, lights, etc. To simulate this environment, the TEGs were fixed between a heat sink and a thin aluminum plate, which was then attached to a hot plate using thermal grease. The experimental setup is shown in Figure 1.12. In order to monitor the temperature of the hot and cold side of the TEG, Omega CO-1 thermocouples were used. The thermocouple was only 0.13 mm thick so it could be placed bonded to the hot and cold sides of the TEG.

A simple circuit has been constructed to take the electrical energy generated by the two energy harvesting devices and store it in a nickel metal hydride battery. The circuit used in this study is shown in Figure 1.13. The diode is a necessary part of this circuit because it forces current to flow only in one direction. If the diode were not present, the TEG would draw power form the battery during times when the voltage generated was less than the voltage of the battery, or, if the hot and cold sides of the TEG switch, a negative voltage will be applied to the battery, causing it to discharge. The output of the TEG is a DC signal, so it does not require a means of rectifying, which is a source of energy loss in piezoelectric power harvesting.

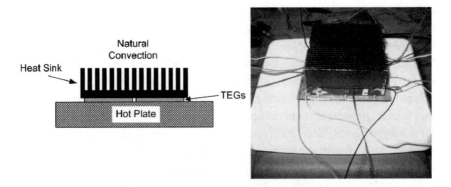

Figure 1.12   Experimental setup of the energy harvesting system

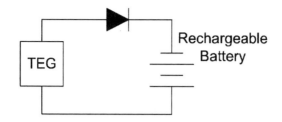

Figure 1.13   Diagram of circuit used to recharge batteries

Table 1.4   Charge times of TEG with solar energy and waste heat, and for a piezoelectric material experiencing random vibration

Battery size	Charge time from solar energy	Charge time from waste heat	Charge time from vibration using piezoelectric materials
80 mAh	3.3 min	NA	2 h
300 mAh	17.3 min	3.5 min	5.8 h

After identifying the ability to use a TEG for the purpose of recharging small batteries, the results were compared to those found using piezoelectric materials by Sodano *et al.* (2007). The time required for each system to charge the battery to a cell voltage of 1.2 V was measured in each case and the results are provided in Table 1.4. The results from charging a battery using piezoelectric materials are only provided to give an idea of the results from previous studies performed under realistic conditions and do not mean that a TEG will always outperform piezoelectric materials. The charge times in Table 1.3 may seem to be lower than possible; the time listed is simply to achieve the batteries' cell voltage and not a full charge. The time needed to take the battery up to capacity would be longer; however, the times listed for both the piezoelectric and the TEGs represent the time required to reach the cell voltage from a fully discharged state. In order to determine the time needed to provide a complete charge of the battery, a charge controller would be needed. The superior performance of the TEG is due to its large current output, whereas the piezoelectric material supplies a very high voltage at a low current. To give an idea of the difference in these devices, the impedance of one TEG is approximately 3 ohms while the piezoelectric impedance is approximately 10 000 ohms. Due to the lower TEG impedance, eight modules had to be connected electrically in series to boost the output voltage to the required 1.2 V of the battery; however, this lower impedance also

Table 1.5 Time required to charge different-sized batteries using a piezoelectric

Battery size (mAh)	Time for charge at resonance (h)	Time for charge with random signal (h)
40	0.62	1.6
80	1.2	2
200	4	1.2
300	6	5.8
750	7	8.6
1000	22	32

makes the TEG far more suited for use with rechargeable batteries, which charge faster with larger currents.

Next, we consider some basic results in harvesting ambient vibration energy from random background vibration such as found near machinery. These are shown in Table 1.5 taken from Sodano *et al.* (2005a).

1.7.2 Vibration Harvesting with Piezoceramics

The piezoelectric effect exists in two domains: the first is the direct piezo-electric effect that describes the material's ability to transform mechanical strain into electrical charge, and the second form is the converse effect, which is the ability to convert an applied electrical potential into mechanical strain energy. The direct piezoelectric effect is responsible for the material's ability to function as a sensor and the converse piezoelectric effect is accountable for its ability to function as an actuator. It is the sensor, or direct, effect that allows piezoelectric material to be used in energy harvesting. A piezoceramic material that is strained, through vibration for example, produces an electric charge which can be bled off and used to produce a voltage and current. Thus piezoceramic materials provide a mechanism to harvest mechanical energy, change it to electrical energy and use it for something else.

Piezoelectric materials belong to a larger class of materials called ferroelectrics. One of the defining traits of a ferroelectric material is that the molecular structure is oriented such that the material exhibits a local charge separation, know as an electric dipole. Throughout the material composition, the electric dipoles are orientated randomly, but when the material is heated above a certain point, the Curie temperature, and a very strong electric field is applied, the electric dipoles reorient themselves relative to the electric field. This process is termed poling. Once the material is cooled, the dipoles maintain their orientation and the material is then said to be poled. After the poling process is completed, the material will exhibit the piezoelectric effect.

After the material has been poled, an electric field can be applied in order to induce an expansion or contraction of the material. However, the electric field can be applied along any surface of the material, each resulting in a potentially different stress and strain generation. Therefore, the piezoelectric properties must contain a sign convention to facilitate this ability to apply electrical potential in three directions. For the sake of keeping this discussion simple, the piezoelectric material can be generalized for two cases. The first is the stack configuration that operates in the '33' mode and the second is the bender, which operates in the '13' mode. The sign convention assumes that the poling direction is always in the '3' direction. With this point the two modes of operation can be understood. In the '33' mode, the electric field is applied in the '3' direction and the material is strained in the poling or '3' direction; in the '13' mode, the electric field is applied in the '3' direction and the material is strained in the '1' direction or perpendicular to the poling direction. These two modes of operation are particularly important when defining the electromechanical coupling coefficient that occurs in two forms. The first form is the actuation coefficient d, and the second is the sensor coefficient g. Thus, g_{13} refers to the sensing coefficient for a bending element poled in the '1' direction and strained along '3'.

Here we examine some simple lab-based energy harvesting experiments to learn a little about the nature of harvesting using piezoelectric devices. These tests compare the use of monolithic piezoceramics, operating in the g_{13} mode, and active fiber composites, operating in the g_{33} mode. The expectation was that the lighter composites would produce more useable energy than the monolithic piezoceramics because of their higher coupling coefficient. However, this was not the case.

Active fiber composites are layered devices essentially consisting of piezoceramic fibers encased in Kapton and covered with a grid of electrodes (called interdigitated), which results it the g_{33} mode of the piezoceramic being activated when strained. There are two commercially available devices made this way: active fiber composites (AFCs) and macro fiber composites (MFCs), each manufactured in different ways. Here we compare the use of MFCs to monolithic piezoceramics (PZT) in terms of their ability to harvest vibration energy. The experiment consisted of mounting similar-sized MFC and PZT in a cantilever position off of a shaker, exciting the shaker with a random signal measured from a compressor, and running the output of the piezoelectric devices through a bridge circuit for battery charging.

First, an efficiency of a simple harvesting system is defined. To experimentally determine efficiency a laser vibrometer is used to measure the displacement of a plate covered with a piezoelectric material and a force transducer to measure the applied force. With this data, and the voltage output from the piezoelectric material, the average efficiency η was numerically calculated from

$$\eta = \frac{P_{out}}{P_{in}} \times 100\% = \frac{\sum_{n=2}^{m} \frac{(V_n - V_{n-1})^2 / R}{((F_n - F_{n-1}) \cdot (d_n - d_{n-1})) / (t_n - t_{n-1})}}{m} \times 100\%$$

Here η is the efficiency, V is the voltage drop across resistance R, F is the force applied to the base of the plate, d is the displacement of the plate, t is the time increment between data points, n is the data point index and m is the highest measured point. The efficiency of three input signals was calculated with the input signals being resonance, chirp and random. The resulting efficiencies are shown in Table 1.6. For each signal three measurements were made to show consistency. The efficiency of the PZT plate is low at resonance because the resonance frequency used was the frequency at which the voltage output was the highest, not the frequency with the best force into voltage out characteristics. This lower efficiency is shown because the resonance frequency is used to charge a battery.

While the MFC had a voltage far larger than the PZT, the power produced was much less. The lower power may be due to the construction of the MFC using piezofibers and interdigitated electrodes providing some additional effects. MFCs are constructed in a diced interdigitated fashion, and the

Table 1.6 Efficiency of PZT and MFC with three different inputs (from Sodano *et al.*, 2003)

Signal	PZT efficiency (%)	MFC efficiency (%)
Resonance	1.1675	0.9442
	2.0777	1.0727
	1.1796	0.8782
Chirp 0–500 Hz	3.927	2.7421
	3.9388	2.5476
	3.8948	2.6285
Random 0–500 Hz	3.9369	0.7636
	3.6825	0.828
	4.2174	0.7366

segments of piezoelectric material between each electrode can be considered a small power source. The majority of these small power sources are connected to one another in series. When two power sources are connected in series the voltages add but the current does not. For this reason, the MFC produces a much higher voltage while the current remains far smaller than that of the monolithic PZT. The fiber composite actuators, while promising higher electromechanical coupling than monolithic piezoceramics, are plagued with a low current output, which hinders the rate at which they can charge a battery. This explanation is still under investigation.

The actual results of charging various batteries are given above in Table 1.5. Here the results are only shown for the PZT cases. The important thing to note here is that even with low levels of power, a relatively small amount of random energy can be used to charge up a reasonable-sized battery in a few hours. For the sensing applications we have in mind here, the time required to recharge a battery is well within the useful range for many structures. For instance, if the system of Figure 1.1 is used on a structure that needs to be examined once every 2 h, with the ambient energy required from Table 1.2, then the 200 mAh battery could be used to run the system, take data, compute the damage metric, broadcast the state of health and then put itself to sleep for and hour and a half while the system recharges the battery.

The results here are based on batteries, but one could also use supercapacitors as a storage device. These are compact, lightweight and typically smaller than a battery. The technology is new, but it may just be that supercapacitors will provide an excellent solution to integrate into the system of Figure 1.1.

1.8 AUTONOMOUS SELF-HEALING

SHM can be combined with smart materials to form systems capable of healing themselves once damage has been determined. A device-level example is given here, while a materials approach is given in Chapter 9 of this volume. The basic idea is presented here in terms of a bolted joint having the ability to assess its current preload and, if too loose, to tighten itself up (Peairs et al., 2004b; Antonios et al., 2006). The idea is fairly simple: a bolted joint is monitored using, in this case, the impedance-based SHM system. A correlation is made between the impedance profile and the bolt's preload. The bolted joint is fitted with a shape memory alloy (SMA) washer. When the SHM algorithm predicts that the bolt has lost its preload, the SMA washer is activated, causing it to expand and regain the appropriate preload. A schematic of the concept is given in Figure 1.14.

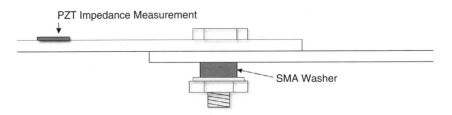

Figure 1.14   A schematic of the self-healing bolted joint concept

SMAs have several distinct features. The shape memory effect is the ability of certain metal alloys to deform as the crystal structure changes from one state (austenitic) at high temperature to another state (martensitic) at low temperature. This phase transition is due to twinning and de-twinning crystal planes. The material deformation occurs as a result of the movement of twin planes without moving dislocations. This allows strains to be easily recovered, and the material appears to remember its shape. Up to 8 % strain can be recovered in this way. Here the SMA is used in constrained recovery where heat is applied to cause it to expand to its original configuration.

Initial experiments were done to show the ability of SMA washers to restore preload in loosened joints. A 0.965 cm thick cylindrical SMA (Nitinol) washer was located between the nut and clamped members, as seen in Figure 1.14. The washer had an inner diameter of 2.436 cm and outer diameter of 2.677 cm.

The impedance response function was measured with the bolt tightened to 40.7 N m, representing its undamaged state. The bolt was then loosened to 13.6 N m to represent damage, and the impedance response function was again measured. Note that while this seems quite a lot, the bolt is still tight, and this amounts to about a quarter turn of the bolt. The entire structure was then heated in an oven to actuate the SMA. The actuator expanded axially (as well as contracting radially) to restore the preload in the joint. A final impedance measurement was then taken. The response of the beam after actuation returns from the damaged state to a state near that of the undamaged condition. After actuation and removal from the joint, the actuator was 0.979 cm thick and had an inner diameter of 2.349 cm and outer diameter of 2.624 cm. These simple experiments motivated sorting out a device to automate the procedure and provide local heating.

SMA can be heated by passing a current through it and inducing resistive heating. Unfortunately, this also tends to heat up all the other metal in the lap joint, including the beams, requiring significant amounts of energy. The solution is to provide a local heater to wrap around the SMA washer and thermally insulate it. With the temperature controlled by a heater, one is able to formulate a model to predict the preload as a function of temperature. This model in turn can help map the impedance measurements to the value of temperature that needs to be applied in order to return the system to its designed preload. This is reported in Antonios *et al.* (2006).

Once this system is designed, the decision to 'fire' the SMA washer can be made in an automated way by continually monitoring plots such as those of Figure 1.7. Once the impedance metric exceeds a predetermined value, a decision circuit can turn on the heater to a specified temperature, regaining the desired preload. This concept illustrates a 'Level 7' system as described in Section 1.2 and forms an example of a 'self-healing system'. This system of Figure 1.14 is self-healing in the sense that it: determines that damage exists; determines how severe the damage is; and then takes action to recover from the damage and restore the system's original function.

1.9 THE WAY FORWARD: AUTONOMIC STRUCTURAL SYSTEMS FOR THREAT MITIGATION

Military and security issues often drive research. In this case, one of the motivators for having autonomous sensing is the desire of several governments to mitigate threats against their military and civilians. The US military's research establishments (Air Force Office of Scientific Research, Army Research Office, and Office of Naval Research) combined with the US National Science Foundation and the European Science Foundation held a workshop in May of 2006 on 'Autonomic Structural Systems for Threat Mitigation'. Autonomic structural systems are loosely defined as systems which respond to external threat in an autonomous way. The workshop's focus was on load-bearing composite material structures with the hope of producing a road map forward for developing multifunctional materials and structures capable of (a) sensing and diagnosing of threats, (b) penetration prevention, (c) load capacity preservation and (d) functionality restoration.

Developing examples of autonomic systems will require the collaboration of engineers and scientists from a variety of disciplines and should be motivated or inspired by looking at biological solutions to threat mitigation (such as bone regrowth). Certain components of the proposed autonomic systems already exist. There are several contributors to this volume who have looked at self-healing, some who have looked at autonomous sensing (as in this chapter and the references) and some who have looked at load-bearing sensors. The concept of multifunctional structures has been around for some time. The import of the proposed autonomic structural and material systems is to bring yet another level of integration together with the hopes of producing structures that autonomously 'take care of themselves' under a variety of threats.

Primitive elements of autonomic systems already exist in the form of airbags and crush zones in automobiles, the wireless sensing and power harvesting systems mentioned above, and the recent work in self-healing materials and structures. The way forward proposes a more sophisticated level of integration and autonomy. Components of embedded neural computing and multifunctional sensing are proposed, with innovations demanded in three areas:

1. Multifunctional sensing and actuation
2. Integrated sensing, computing, informatics and communication
3. Predictive and proactive sensing

Multifunctional sensing refers to the concept of a mobile, multidirectional sensor that could respond to a variety of length and time scales, change its sensing mechanism (say from acoustic pressure waves to chemical), change its orientation and reprogram its function. Actuation refers to mechanical, chemical and electromagnetic force. Actuation and sensing can be combined to form the concept of a *morphing sensor* that could change its physical form to adjust to the appropriate threat. These ideas are inspired by the many living creatures on Earth and the way in which our own bodies respond to threats and/or damage.

While these ideas may seem far fetched by today's standards, they motivate the scientific and technological issues of (a) expanding the 'abilities' of our current sensing systems and materials, (b) integrating sensor functions, actuation and intelligence (i.e. computing), (c) expanding current functionality and levels of integration, and (d) improving our energy harvesting and power management systems.

Our own particular efforts in these directions focus on continuing to integrate multiple energy harvesting methods into a single load-bearing structure. We are further pursuing chip computing without an operating system to minimize the amount of power needed to run algorithms at the sensor location (integrated computing). On the harvesting front, we hope to examine a particular application of a micro air vehicle to extend its range and performance by integrating multiple harvesting sources into structural components and focusing on video sensing. In addition, we hope to miniaturize our current prototype described above to provide a more highly integrated structural sensor that is completely autonomous. While these are minor perturbations of our existing systems in the context of the way forward to an autonomous system, they form necessary steps to eventually achieving an autonomic structure.

1.10 SUMMARY

This chapter presents the first fully self-contained system that performs impedance-based SHM. In previous research, a system was developed which performed most of the health monitoring steps, but needed the use of an external function generator for actuation. The current autonomous system effectively replaces an impedance analyzer and external data analysis. All of the structural excitation, data acquisition and health monitoring analysis are performed in a matter of seconds. With traditional impedance techniques, after the data is acquired, all of the analysis must still be done using processing software to determine whether there is damage. Now, damage in a structure can be found almost immediately.

Also described is the first use of impedance excitation with targeted sinc functions. The use of sinc functions has the potential to save both excitation time and computational power. By slightly varying the fundamental frequency with each pulse, the structure will be excited at every frequency in the range of interest.

A quick review of methods for capturing ambient vibration and thermal energy for use in for battery charging is also presented. The idea proposed here is to produce a completely wireless system capable of providing autonomous structural health monitoring.

A simple example of a self-healing bolted joint is also presented as an illustration of an autonomous self-monitoring and healing system. All of this leads to the future direction of adaptive, autonomous structures as described above. In the short term, future work on the autonomous system includes performing a complete excitation signal energy analysis to explore the benefits of actuation with sinc functions. Also, piezoelectric-based and thermal power harvesting will be incorporated to allow the system to be fully self-sufficient. Eventually, with the knowledge gained from this prototype, an even smaller prototype can be custom designed with components specific to the project, all leading to the eventual goal of having a complete impedance-based SHM system contained on a single chip.

ACKNOWLEDGEMENTS

The authors would especially like to thank Hyung-Jin Lee, advised by Dr. Dong S. Ha, of the Virginia Tech Electrical and Computer Engineering Department, for all his hard work and help on this project. Dr. Robert Owen of Extreme Diagnostics, Inc. and Dr. Gyuhae Park of Los Alamos National Laboratory also contributed to this research. The authors gratefully acknowledge funding for this research provided by NASA Langley Research Center and Extreme Diagnostics, Inc. (contract NNL05AB08P). This material is based in part upon work supported by the National Science Foundation under Grant No. 0426777.

REFERENCES

Antonios, C., Inman, D.J. and Smaili, A., 2006, 'Experimental and Theoretical Behavior of Self-healing Bolted Joints,' *Journal of Intelligent Material Systems and Structures*, **17**(6), pp. 499–509.

Doebling, S.W., Farrar, C.R. and Prime, M.B., 1998, 'A Summary Review of Vibration-Based Damage Identification Methods,' *Shock and Vibration Digest*, **30**(2), pp. 91–105.

Dosch, J.J., Inman, D.J. and Mayne, R.W., 1994, 'Dual Function System Having a Piezoelectric Element (Self Sensing Actuator),' US Patent #5,347,870 awarded September 20, 1994.

Giurgiutiu, V. and Zagrai, A.N., 2002, 'Embedded Self-Sensing Piezoelectric Active Sensors for On-line Structural Identification,' *Journal of Vibration and Acoustics*, **124**, pp. 116–125.

Grisso, B.L., Martin, L.A. and Inman, D.J., 2005, 'A Wireless Active Sensing System for Impedance based Structural Health Monitoring', *Proceedings of IMAC XXIII*, Orlando, FL, January 31–February 3.

Guyamor, D., Badel, D., Leveuvre, E. and Richard, C., 2005, 'Toward Energy Harvesting using Active Materials and Conversion Improvement by Nonlinear Processing', *IEEE Transactions on Ultrasonics, Ferroelectrics and Frequency Control*, **53**(4), pp. 584–596.

Inman, D.J., Farrar, C.R., Steffan, V. and Lopez, V., Ed., 2005, *Damage Prognosis*, John Wiley and Sons, Ltd, Chichester.

Liang, C., Sun, F.P. and Rogers, C.A., 1994, 'An Impedance Method for Dynamic Analysis of Active Material System,' *Journal of Vibration and Acoustics*, **116**, 121–128.

Lynch, J.P. and Loh, K., 2005, 'A Summary Review of Wireless Sensors and Sensor Networks for Structural Health Monitoring,' *Shock and Vibration Digest*, **38**(2), pp. 91–128.

Lynch, J.P., Sundararajan, A., Law, K.H., Kiremidjian, A.S., Carryer, E., Sohn, H. and Farrar, C.R., 2003, 'Field Validation of a Wireless Structural Monitoring System on the Alamosa Canyon Bridge,' *SPIE's 10th Annual International Symposium on Smart Structures and Materials*, San Diego, CA.

Lynch, J.P., Sundararajan, A., Law, K.H., Kiremidjian, A.S., Kenny, T.W. and Carryer, E., 2002, 'Embedment of Structural Monitoring Algorithms in a Wireless Sensing Unit,' *Structural Engineering and Mechanics*, **15**(3), pp. 285–297.

Lynch, J.P., Sundararajan, A., Law, K.H., Kiremidjian, A.S. and Carryer, E., 2004a, 'Embedding Damage Detection Algorithms in a Wireless Sensing Unit for Attainment of Operational Power Efficiency.' *Smart Materials and Structures*, **13**, pp. 800–810.

Lynch, J.P., Sundararajan, A., Law, K.H., Sohn, H. and Farrar, C.R., 2004b, 'Piezoelectric Structural Excitation Using a Wireless Active Sensing Unit', *Proceedings of the 22nd International Modal Analysis Conference*, on CD by the Society of Experimental Mechanics.

Mascarenas, D.L., Todd, M.D., Park, G. and Farrar, C.R., 2006, 'A Miniaturized Electromechanical Impedance-based Node for the Wireless Interrogation of Structural Health', *Proceeding of SPIE's 13th Annual International Symposium on Smart Structures and Materials*, **6177**, March 28.

Park, G., Sohn, H., Farrar, C.R. and Inman, D.J., 2003, 'Overview of Piezoelectric Impedance-Based Health Monitoring and Path Forward,' *Shock and Vibration Digest*, **35**(6), 451–463.

Peairs, D.M., Park, G. and Inman D.J., 2004a. 'Improving Accessibility of the Impedance-based Structural Health Monitoring Method,' *Journal of Intelligent Material Systems and Structures*, **15**(2), 129–140.

Peairs, D.M., Park, G. and Inman, D.J., 2004b, 'Practical Issues of Activating Self-repairing Bolted Joints,' *Smart Materials and Structures*, **13**, pp. 1414–1423.

Radiometrix Ltd., 2005, *UHF Narrow Band FM multi channel radio modules*, Radiometrix Inc., Harrow, Middlesex.

Roundy, S., 2005, 'On the Effectiveness of Vibration-Based Energy Harvesting,' *Journal of Intelligent Material Systems and Structures*, **16**, pp. 809–823.

Roundy, S., Wright, P.K. and Rabaey, J.M. 2004, *Energy Scavenging for Wireless Sensor Networks: With Special Focus on Vibrations*, Kluwer Academic, Dordrecht.

Schilling, R.J. and Harris, S.L., 2005, *Fundamentals of Digital Signal Processing using MATLAB*, Nelson, Toronto.

Sodano, H.A., Park, G., Leo, D.J. and Inman, D.J., 2003, 'Use of Piezoelectric Energy Harvesting Devices for Charging Batteries,' *Proceeding of SPIE's 10th Annual International Symposium on Smart Structures and Materials*, **5050**, pp. 101–108, July.

Sodano, H.A, Park, G. and Inman, D.J., 2004a, 'Estimation of Electric Charge Output for Piezoelectric Energy Harvesting,' *Journal of Strain*, **40**, pp. 49–58.

Sodano, H., Park, G. and Inman, D.J., 2004b. 'A Review of Power Harvesting from Vibration using Piezoelectric Materials,' *Shock and Vibration Digest*, **36**(3), pp. 197–205.

Sodano, H.A., Park, G. and Inman, D.J., 2005a, 'Comparison of Piezoelectric Energy Harvesting Devices for Recharging Batteries,' *Journal of Intelligent Material Systems and Structures*, **16**(10), pp. 799–808.

Sodano, H.A., Inman, D.J. and Park, G., 2005b. 'Generation and Storage of Electricity from Power Harvesting Devices,' Journal *of Intelligent Material Systems and Structures*, **16**(1), pp. 67–75.

Sodano, H.A., Simmers, G.E., Dereux, R. and Inman, D.J., 2007, 'Recharging Batteries Using Energy Harvested from Thermal Gradients,' *Journal of Intelligent Material Systems and Structures*, **18**(1), pp. 3–10.

Sohn, H., Allen, D.W., Worden, K. and Farrar, C.R., 2003, 'Statistical Damage Classification Using Sequential Probability Ratio Tests,' *An International Journal of Structural Health Monitoring*, **2**, pp. 57–74.

Sohn, H., Worden, K. and Farrar, C.R., 2002, 'Statistical Damage Classification Under Changing Environmental and Operational Conditions,' *Journal of Intelligent Material Systems and Structures*, **13**, pp. 561–574.

Spencer, Jr., B.F., Ruiz-Sandoval, M.E. and Kurata, N., 2004, 'Smart Sensing Technology: Opportunities and Challenges,' *Journal of Structural Control and Health Monitoring*, **11**(4), 349–368.

Straser, E.G. and Kiremidjian, A.S., 1998, *A modular, wireless damage monitoring system for structures*. Report No. 128, John A. Blume Earthquake Engineering Center, Department of Civil and Environmental Engineering, Stanford University, Stanford, CA.

Tanner, N.A, Wait, J.R., Farrar, C.R. and Sohn, H., 2003, 'Structural Health Monitoring Using Wireless Sensing Systems with Embedded Processing', *Journal of Intelligent Materials Systems and Structures*, **14**(1), pp. 43–55.

Texas Instruments, 2005a, *Code Composer Studio Development Tools v3.1 Getting Started Guide*, Texas Instruments Inc., Dallas, Texas.

Texas Instruments, 2005b, *TMS320C6713, TMS320C6713B Floating-Point Digital Signal Processors*, Texas Instruments Inc., Dallas, Texas.

Texas Instruments, 2002, *ADS8364EVM User's Guide*, Texas Instruments Inc., Dallas, Texas.

Texas Instruments, 2001, *TLV5619-5639 12-Bit Parallel DAC Evaluation Module User's Guide*, Texas Instruments Inc., Dallas, Texas.

2

Distributed Sensing for Active Control

Suk-Min Moon, Leslie P. Fowler and Robert L. Clark

Duke University, Department of Mechanical Engineering and Materials Science, Box 90300, Durham, NC 27708-0300, USA

2.1 INTRODUCTION

The level of performance achieved in adaptive structures devoted to active control of sound and/or vibration is critically dependent upon the sensors used in the feedback and/or feedforward control paths. A block diagram of the typical two-input, two-output system, where the inputs and outputs can be scalars or vectors, is depicted in Figure 2.1. The sensor(s) ($y(s)$ in Figure 2.1) must provide coherent signals with the exogenous input signals that define the disturbances ($w(s)$ in Figure 2.1), and the sensor(s) should be selected so as to emphasize only the plant dynamics important in the desired performance path of the system ($G_{zw}(s)$ in Figure 2.1). Ideally, if a control path can be identified that mirrors the dynamics of the performance path (i.e., $G_{yu}(s) = G_{zw}(s)$), then all of the control energy can be devoted to the most important structural and/or acoustic modes present in that path. In practice this is typically not the case, and more frequent than not, the specific disturbance path is unknown.

A significant body of research (more than can be practically cited here) has been devoted to optimization of actuator and sensor aperture and placement

---

Adaptive Structures: Engineering Applications Edited by D. Wagg, I. Bond, P. Weaver and M. Friswell

Figure 2.1 Block diagram of model system

[8–9, 11–12, 13, 15]. Performance metrics and optimization routines have been developed that focus on actuators and sensors individually and simultaneously. In most cases, the resulting design is limited primarily by the designer's knowledge of the system and the chosen performance metric.

Additionally, a significant body of research has evolved in principal component analysis, a mathematical tool which has found broad application in signal processing and control theory through application of singular value decomposition (SVD) [1–5]. In signal processing, principal component analysis is frequently used to determine the number of independent sources present in an array of signals, often obtained from an array of sensors. Additionally, SVD has been utilized in control theory for applications ranging from model order reduction to the development of performance metrics for actuator and sensor placement [8, 9, 12, 13, 15, 19].

Within this chapter, an approach for estimating the number of exogenous disturbance sources in a structural system is investigated. The results presented demonstrate that the application of feedback control can be used to identify additional disturbances that are initially 'buried' by the magnitude of the dominant disturbance source. If the objective is to identify sensors that convey multiple disturbances and those present over more than an order of magnitude, the use of feedback control to suppress the dominant source can facilitate identification of additional disturbance sources.

Once an appropriate array of sensors capable of capturing the plant response to exogenous disturbances has been identified, a mechanism for optimizing this array so as to minimize the number of inputs to the control system is desired. To this end, a method for finding an optimal subset of sensors from a given array or a weighted output of the array that conveys the dynamics important in the performance path is presented. As was demonstrated previously, it is possible to optimize the selection of sensors that couple to targeted modes represented in the performance metric and ignore those modes unimportant for control over the targeted bandwidth [12, 20–22]. Results from experimental implementation of a real-time sensor selection/weighting algorithm demonstrate that it is feasible to adapt (a) the temporal compensation achieved through the application of adaptive generalized predictive control [10] and (b) the spatial compensation achieved through design optimization based upon real-time system identification in parallel to enhance performance for varying disturbance inputs. Basic design procedures are outlined for simultaneous spatial and temporal compensation, and both analytical and experimental models are used to demonstrate the approaches.

2.2 DESCRIPTION OF EXPERIMENTAL TEST BED

The experimental test bed developed for this study includes both optical components and a flexible structure. The objective was to construct an experimental system with many degrees of freedom and unknown boundary conditions so as to provide a greater level of complexity than that typically characterized by 'academic' structures. The flexible panel is configured with a mounting post and turning mirror that are part of the optical path and a number of proof-mass actuators which can be used to provide exogenous vibratory disturbance inputs to the structure.

The experimental test bed is shown in Figure 2.2. A laser source and a position sensing detector (PSD) are mounted on one optical bench. The optical path is completed when the light emanating from the lasers returns to the PSD. The PSD is used to quantify optical jitter resulting from disturbances introduced over the optical path. A fast steering mirror (FSM) is mounted on another optical bench and is used as the control input to minimize jitter.

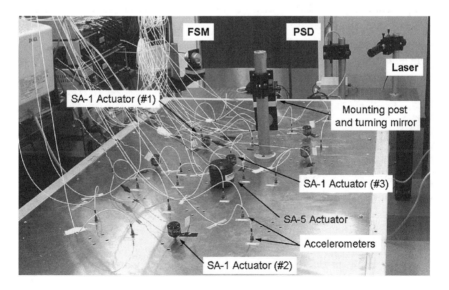

Figure 2.2   The experimental test bed

A flat turning mirror and its mounting post are mounted on an aluminum plate 3 feet by 5 feet by 1/8 inch thick (approximately 1 metre by 1.6 metres by 0.3 centimetres). The aluminum plate is bolted to a wooden frame and placed on a lab table for alignment with the other optical benches.

In addition to the optical components, proof-mass actuators (Model #: SA-1, CSA Engineering, Inc.) and accelerometers (Model #: D352C65, PCB Piezotronics) are mounted on the surface of the flexible plate. The proof-mass actuators are used as disturbance sources for the plate and thus create jitter in the optical path through vibratory motion induced in the turning mirror through its post, which is mounted on the plate. The accelerometers are used to measure the plate vibration.

2.3   DISTURBANCE ESTIMATION

The initial phase of the study focused on disturbance estimation. Here, disturbance estimation is restricted to simply estimating the number of independent disturbances present in the dynamic system.

2.3.1   Principal Component Analysis

The proposed technique for estimating the number of uncorrelated disturbance sources applied to the system from sampled data is based on principal component analysis (PCA) [1–5]. PCA provides a mechanism for identifying patterns in data and expressing the data in such a way as to highlight their similarities and differences.

In general terms, when an m-dimensional data set, $\mathbf{y}$, is available, an m by m covariance matrix, $\mathbf{R}$, can be calculated,

$$\mathbf{R} = \text{cov}(\mathbf{y}_0) \tag{2.1}$$

where $\mathbf{y}_0$ is the data set with zero mean, obtained by subtracting the mean from each of the data dimensions.

The SVD of the covariance matrix gives the singular values and singular vectors of the covariance matrix,

$$\mathbf{R} = \mathbf{U}^*\mathbf{S}^*\mathbf{V}' \tag{2.2}$$

where an m by m diagonal matrix, $\mathbf{S}$, has singular values on its diagonal and $\mathbf{U}$ and $\mathbf{V}$ are the left and right singular vectors respectively. Note that for a real, symmetric matrix such as the covariance matrix, the left and right singular vectors are the same ($\mathbf{U} = \mathbf{V}$). For an m-dimensional data set, there exist m singular values and m corresponding singular vectors.

When the largest singular value is much greater than the second-largest singular value, it is said that the data set is related along the corresponding singular vector to the largest singular value [1, 4, 5].

Based upon this observation, a weighted output can be used to estimate the response in the principal coordinate. Here, the principal response, $\mathbf{y}_{pc}$, is defined as the weighted sum of sensor signals by the singular vector corresponding to the largest singular value:

$$\mathbf{y}_{pc} = \mathbf{y} \cdot \mathbf{V}_1 \qquad (2.3)$$

where $\mathbf{y}$ is a sensor signal vector, and the principal singular vector, $\mathbf{V}_1$, is the singular vector corresponding to the largest singular value. This provides a single measured output for application in control that is a linear combination of all sensors used in the array.

2.3.2 Application of PCA: Case Studies

To demonstrate application of the PCA/SVD method, a simple analytical study was conducted first. The study was followed with an experimental demonstration of the method. It should be noted that this study demonstrates that the standard PCA/SVD method is insufficient in determining the correct number of relevant exogenous disturbance inputs. Secondary disturbance sources, which may well be an order of magnitude below the primary disturbance at the source or as measured by the reference sensors due to dynamics along the structural path, are not well observed and thus may limit overall performance in control.

2.3.2.1 Analytical Example

Consider the multiple degree-of-freedom, mass–spring–damper system model ($m_i = 10\,\text{kg}$, $k_i = 100\,\text{N/m}$, and $c_i = 0.1\,\text{N} \cdot \text{s/m}$) shown in Figure 2.3. For the initial test case, uncorrelated stochastic force signals were applied to the 5th and 6th masses, which will be designated Model 1. In the second test case, uncorrelated stochastic force signals were applied to the 5th and 20th masses, which will be designated Model 2. For each test model, time-domain acceleration measurements were obtained at masses 1 through 10. The singular values of the covariance matrix corresponding to these time-domain signals were computed. When the disturbance sources and sensors

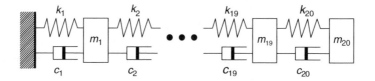

Figure 2.3   Analytical mass–spring–damper system

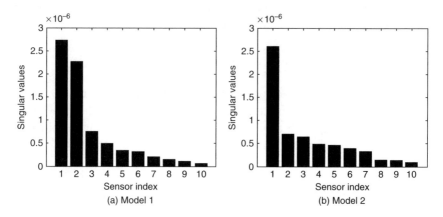

Figure 2.4   Singular value plot of mass–spring–damper system

are proximate to each other as in Model 1, the number of dominant singular values equals the number of disturbance sources as shown in Figure 2.4(a). However, for Model 2, only one dominant singular value is observed in Figure 2.4(b). Note that the term dominant singular value(s) is used when the difference between the largest singular value(s) and the second-largest singular value(s) is much greater than the differences of any other adjacent singular values (singular value indices, i and $i + 1$, where $i > 2$) [1, 4]. The reason that the PCA method yields only one dominant singular value in Model 2 is that the sensor response due the force input at mass 5 is much larger (as expected) than the response due to the force input at mass 20. Hence, the exogenous disturbance source at mass 20 does not influence the singular value plot.

2.3.2.2 Experimental Example

To demonstrate the results experimentally, acceleration data was collected from the plate structure described in the previous section and illustrated in Figure 2.2. A total of 23 accelerometer sensors were used to measure the structural response. Uncorrelated band-limited (0–2000 Hz) stochastic inputs were amplified and applied to the actuators used in each test case.

For the first experiment, a band-limited stochastic input was applied to one of the actuators mounted on the plate structure (SA-1 actuator #1 in Figure 2.2). Acceleration data from all 23 accelerometers was collected and the covariance matrix was computed. The corresponding singular values were computed and are plotted in Figure 2.5(a). A similar experiment was repeated with another actuator on the structure (SA-1 actuator #2 in Figure 2.2). The singular values of the covariance matrix for the corresponding data are also plotted in Figure 2.5(a). The largest singular value computed for the test using actuator #1 is two times greater than the largest singular value computed for the test using actuator #2. From the results obtained, the singular values are observed to decrease with a low-order 'decay'. As detailed previously, most metrics for selecting a cut-off criterion for the number of identified independent disturbance sources do so by comparing successive differences in the singular values. When the difference between singular values is small compared to the previous difference, the remaining smaller singular values are ignored.

In the second test case, the plate structure is disturbed by both SA-1 actuators #1 and #2 simultaneously with independent, band-limited, stochastic inputs. Figure 2.5(b) shows the singular values of the covariance matrix computed using data from all accelerometers. Although there are two distinct signal sources resulting from two SA-1 actuators, #1 and #2, the singular values obtained differ very little from that of SA-1 actuator #1 acting alone. Thus, the impact of the second actuator is clearly 'hidden' in the SVD

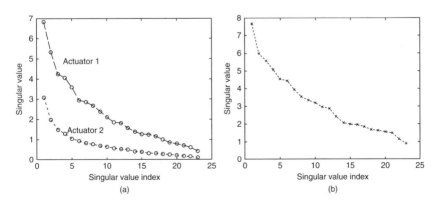

Figure 2.5   Singular value plot of data covariance matrix when signal source(s) is (a) actuators applied separately, and (b) both actuators applied simultaneously

plot, particularly in comparing the SVD plot of actuator #1 acting alone in Figure 2.5(a) to that of both actuators acting together in Figure 2.5(b).

It should be noted that plant dynamics also affect the SVD analysis. In particular, for stochastic inputs, the magnitude of the response will be path dependent and will thus require more than one sensor to capture the full spectrum of the disturbance input.

2.3.2.3 Summary

If more than one disturbance source is present, the response due to each disturbance must be of the same order of magnitude in the measured response to be clearly identified by PCA. Proximity of the disturbances to error sensors or relative gain can affect the number of independent disturbances estimated by PCA. Knowledge of additional disturbances can alter the design and selection of sensor arrays for control and may be important depending upon the desired level performance.

2.3.3 Combining Active Control and PCA to Identify Secondary Disturbances

As indicated in the analytical and experimental examples provided, resolving the number of independent disturbance sources can be challenging when the magnitude of the input differs by an order of magnitude or more. An order of magnitude is significant, and for some disturbance rejection applications, the presence of a secondary disturbance an order of magnitude below that of a primary disturbance may be deemed unimportant. However, if enhanced performance is desired and sufficient control authority is available, estimation of the number of additional disturbances and an appropriate mechanism for sensing them is required.

In this section, a technique is investigated to better estimate the presence of secondary disturbance sources. A secondary disturbance source is defined here as a disturbance source that is not apparent (or difficult to identify) in the singular value plot obtained from a data covariance matrix. To estimate the secondary disturbance sources, an additional control source is used to minimize the system response due to the dominant disturbance source(s). Once the output response resulting from the dominant disturbance source is reduced, the dynamic system response due to the secondary disturbance sources can be observed in the singular value plot.

To accomplish this goal, the output response to be minimized in the control path is the weighted sensor signal produced by multiplying the corresponding singular vector of the principal/dominant singular value by the sensor array as described in Equation (2.3). The resulting controller is thus a single-input, single-output compensator since the weighted array produces one output signal.

2.3.3.1 Analytical Example

A system with a total of 30 mass–spring–dampers was considered. Uncorrelated random force signals were applied to masses 2, 29, and 30 (expanding on the concept depicted in Figure 2.3). Acceleration was measured at masses 1 through 7. Figure 2.6(a) shows the singular value ratio plot of the sensor output covariance matrix. Only one dominant singular value is observed. Obviously this was 'contrived' by placing two of the disturbance sources far from the measured response.

In order to estimate the number of secondary disturbance sources, a fixed gain controller based on the generalized predictive control algorithm was designed [6–7] to minimize the principal response defined in Equation (2.3). (Note that the choice of controller was one of convenience, and alternative control system designs could be considered.) The control input source is arbitrarily selected to be located on mass 10. Upon applying the control

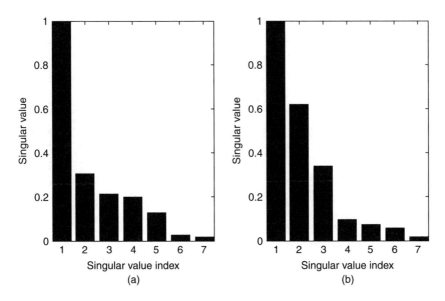

Figure 2.6 Singular value ratio plot: (a) open-loop response, (b) closed-loop response

input, the time-domain sensor output is measured and stored. Figure 2.6(b) shows the singular value ratio plot of the sensor output covariance matrix. As illustrated in Figure 2.6(b), three disturbance sources are now observed: one of the singular values is caused by the control source and the other two singular values are caused by the secondary disturbance sources. Although there are four sources presents in the system – three unknown disturbance sources and one control source – the dominant disturbance source does not appear in the singular value ratio plot because the response corresponding to the dominant disturbance source is now minimized by the control source.

2.3.3.2 Experimental Example

The technique was applied to the experimental plate structure illustrated in Figure 2.2. Three proof-mass actuators were used to disturb the plate structure, and uncorrelated band-limited (0–2000 Hz) stochastic inputs were applied to the actuators. Uniform gain was applied to all disturbance sources; however, the effective gain resulting from structural dynamics and proximity of the disturbance source to the sensor array affects the contribution to the measured response. A covariance matrix was computed from the data collected from all 23 accelerometers and the corresponding singular values are plotted in Figure 2.7(a). As many as two signal source are observed.

An additional proof-mass actuator mounted on the plate structure was then used as a control source, and a fixed-gain generalized predictive controller was designed to minimize the principal response which is obtained by weighting the sensor array outputs by the singular vector corresponding to

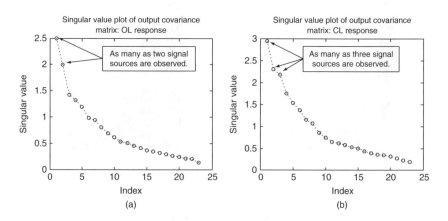

Figure 2.7 Singular value plot: (a) open-loop response, (b) closed-loop response

the largest singular value of the covariance matrix [1–3]. Figure 2.7(b) shows the singular value plot of the output covariance in the presence of the control input. Now as many as three dominant sources are observed in the plot, with two very nearly the same order of magnitude.

2.3.3.3 Summary

The application of active control in combination with PCA has been demonstrated as a useful tool in estimating the number of disturbance sources present from an array of sensors. For disturbance rejection applications aimed at pushing the limits of performance, such information is critical in the overall design.

2.4 SENSOR SELECTION

Once the number of disturbance sources is known and performance objectives have been identified, the task of designing and/or selecting specific sensors must be addressed. An overall block diagram of the system considered in this work is illustrated in Figure 2.8. The plant can be represented by a model or an actual system such as the combined optical/structural system illustrated in Figure 2.2. Disturbances to the system can result from exogenous external sources which may originate from acoustic loading or structural vibrations. The optical/structural system incorporates a fast steering mirror as a secondary actuator to reduce the impact of acoustic and structure-borne disturbances in the optical path and a laser position-sensing detector (PSD) as an error sensor to measure the impact of the disturbances in the optical path. Vibration measurement sensors (accelerometers) are used as reference sensors in the adaptive controller. The overall control objective is to minimize some norm, for example the H_2 norm, of the transfer function from disturbance ω to error sensor y.

As illustrated in Figure 2.8, two sensor selection/design methods were considered:

1. **Method 1**: In this method, the reference sensors were ranked by their ability to couple to targeted mode(s) important in the control bandwidth, and the sensor(s) ranked highest by this metric were selected as error sensor(s) in the control path (see Figure 2.8(a)).

2. **Method 2**: In this method, both sensor pairing and weighting were optimized to produce a single sensor output that couples well to the targeted

mode(s) important in the control bandwidth, and the weighted pair with greatest coupling was selected as the error sensor in the control path (see Figure 2.8(b)).

Once reference sensors were determined by both Method 1 and Method 2, a generalized predictive controller (GPC) was implemented [6, 10, 17–18]. The selection/design methodology was combined with an adaptive control algorithm and all processes were performed in real time. This allows for modification of sensor selection/weighting as well as temporal compensation

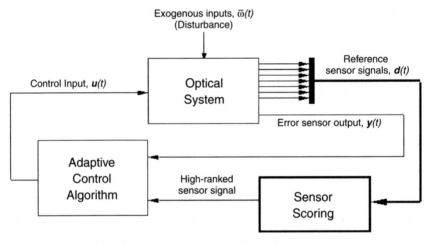

(a) Block diagram of adaptive control algorithm using high-ranked reference sensor

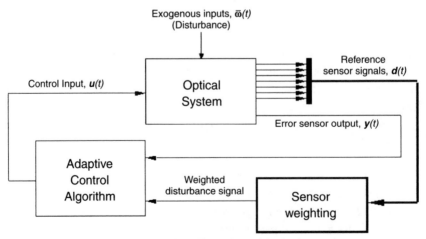

(b) Block diagram of adaptive control algorithm using weighted reference sensor

Figure 2.8   Overall block diagram of closed-loop system

as a function of changes in environmental operating conditions, disturbance paths, and/or system dynamics.

2.4.1 Model Estimation

To integrate a sensor design optimization process into an adaptive control system, the system behavior must be determined through online identification, which refers to determination of the system behavior within a class of mathematical models by means of real-time digital signal processing. For this purpose, the recursive least-squares (RLS) identification algorithm was applied to express the current response of the error sensor signal as a linear function of past response of the error sensor signal, $y(t-i)$, the control input signal, $u(t-i)$, and the reference sensor signals, $d(t-i)$ [7]. By combining $u(t)$ and $d(t)$ in vector form, $v(t) = [u(t) \quad d(t)]^T$, the current response of the error sensor signal can be expressed as

$$y(t) = \sum_{i=1}^{p} \alpha_i y(t-i) + \sum_{i=1}^{p} \eta_i v(t-i) \qquad (2.4)$$

where α_i and η_i are the model parameters to be estimated using the RLS algorithm.

For the test bed used in this work, the error sensor is the PSD, the control actuator is the fast steering mirror, and the reference sensors are the accelerometers. The PSD is used in the feedback path of the GPC and the accelerometers are used in the feedforward path of the GPC. This is possible because the control input is applied through the fast steering mirror, which is physically isolated from the flexible panel where the accelerometers are placed (see Figure 2.2), meaning that the resulting signals from the accelerometers serve as uncontrollable reference signals [11].

One property of the RLS algorithm is that it allows us to omit some data during operation. In the discrete-time application, for instance, if the data sample rate is too fast to perform all computations during a data sampling period, the recursion can be applied at a slower sampling rate. This is permissible because the recursion does not rely on the relation between two successive data histories.

2.4.2 Optimal Sensor Strategy

Once a linear model given by Equation (2.4) is obtained, the reference sensors are ranked by scoring their ability to couple well to the targeted mode(s) for control. For the application discussed here, we seek the accelerometers that best reflect the observed response at the PSD, which serves as an error sensor in feedback control and defines performance in jitter reduction.

2.4.2.1 Sensor Scoring Metric

The basis for a modal coupling scoring metric lies in approximating the Hankel singular values (HSVs) corresponding to system modes for the input/output path of interest. The development of this approximation is based on the assumption that the modes in question are all lightly damped [8, 9, 13, 15]. An approximation of the squared HSVs for the ith mode of the system, γ_i^4, can be written as

$$\gamma_i^4 \approx \frac{tr[\mathbf{B}_i\mathbf{B}_i^T]tr[\mathbf{C}_i\mathbf{C}_i^T]}{16\delta_i^2} \tag{2.5}$$

$$\delta_i = \zeta_i\omega_i \tag{2.6}$$

where the $\mathbf{B}_i$ and $\mathbf{C}_i$ terms are the input and output matrices respectively from the state-space model corresponding to the ith mode. The parameter δ_i is the real part of the natural logarithm of the ith eigenvalue, which as indicated in Equation (2.6) defines the time constant of the mode because it is expressed in terms of the damping ratio, ζ_i, and the natural frequency, ω_i. For the work presented here, the state-space model is obtained from the linear model, given in Equation (2.4), using the eigensystem realization algorithm (ERA) [7, 14].

With the HSV approximations given in Equation (2.5), let Γ_{yd}^2 be the HSV between an error sensor and a reference sensor,

$$\Gamma_{yd}^2 = diag(\gamma_{yd_1}^2, \cdots, \gamma_{yd_n}^2) \tag{2.7}$$

where Γ_{yd}^2 are completely determined from a plant model, P_{yd}, which represents the dynamics between the error sensor and the reference sensor. For the model P_{yd}, the reference sensor signals are considered as inputs and the error sensor is considered an output.

Additionally, let Λ be a binary selection vector with ones corresponding to the targeted modes of the system and zeros elsewhere. This leads to a scoring metric whose values scale according to coupling with selected modes. Furthermore, two possible scoring values can be defined based on the choice of Λ. Letting Λ_c be a binary selection vector for the modes deemed important in the control bandwidth and Λ_d be a vector for the modes unimportant in the control bandwidth, then

$$J_c = \sum_{i=1}^{N} \Lambda_c \gamma_{yd_i}^2 \qquad (2.8)$$

$$J_d = \sum_{i=1}^{N} \Lambda_d \gamma_{yd_i}^2 \qquad (2.9)$$

Two values, obtained from Equations (2.8) and (2.9), yield scores for a pair of error/reference sensors that correspond to their beneficial and unimportant modal coupling (as selected by the designer). In order to compare the scores for each pair of error/reference sensors, the two metrics, J_c and J_d, need to be combined into a single score. One approach is to normalize them by dividing by maximum scores, $\max[J_c]$ for beneficial modal coupling and $\max[J_c]$ for unimportant modal coupling. The final scoring metric is obtained by multiplying the beneficial metric and the reversal of the unimportant metric, i.e.,

$$\mathbf{J} = \bar{J}_c * \bar{J}_d \qquad (2.10)$$

where $\bar{J}_c$ is the normalized score associated with the beneficial modal coupling and

$$\bar{J}_d = \frac{\max[J_d] - J_d}{\max[J_d]} \qquad (2.11)$$

for the unimportant modal coupling.

2.4.2.2 Optimal Sensing Algorithm

Once the RLS algorithm is used to estimate and update the linear system model from measured data, the model is converted to state-space form in order to compute the sensor scoring metric, $\mathbf{J}$, given in Equation (2.10). Based on the computed sensor scoring metric, the highest ranked sensor is selected and its signal is fed through the adaptive control algorithm (see Figure 2.8(a)). It is expected that when there is no significant environmental and/or changes in system dynamics or disturbance paths, the RLS algorithm will converge to its least-squares solution and the sensor index ranking will remain constant.

Furthermore, the sensor scoring metric concepts can be extended to a pair of weighted reference sensors to produce a single combined reference signal.

The applied weighting can be optimized to enhance coupling to modes of interest and reduce coupling to modes unimportant in the control bandwidth. In this study, a genetic optimization algorithm was used in the selection of sensor pairs and their weightings. Once an optimal reference sensor pair is selected, the weighted reference sensor signal is used in the adaptive control design (see Figure 2.8(b)). The following section summarizes the genetic optimization algorithm.

2.4.2.3 Genetic Optimization

The genetic optimization algorithm is designed to converge on an optimal solution through a series of iterations of mating based on the beneficial ranking of the pairs [16]. The definition of an individual pair is defined by two parameters – sensor combination and weighting of each sensor. These parameter vectors are then converted into a binary matrix with 16-digit resolution. A random starting population is constructed prior to implementation of the algorithm. Mating is determined based on a binary crossover strategy. Mutations are performed by allowing random adjustments to individual binary digits over a prescribed percentage of the mating events.

In addition, an adjustment is made to retain the highest-scoring pair for the next iteration. This adjustment forces the winner of each successive iteration to be at least as good as the previous one. This technique consistently yields similar results regardless of the parameters of the original population as detailed by Richard and Clark [12].

The resulting best pair of reference sensors and their weighting after a series of iterations are used to compute the weighted reference sensor signal. This weighted reference sensor signal is then used in the feedforward portion of the adaptive control algorithm.

2.4.3 Experimental Demonstration

Each proposed method in Figure 2.8 could be divided into two processes: (a) the process of the reference sensor scoring/weighting and (b) the process of controller design. In the process of the reference sensor scoring/weighting, model estimation, sensor scoring/optimization, and computation of the reference signal for implementation in the controller are performed. For the purpose of this work, an adaptive generalized predictive control algorithm was applied [6, 10]. The adaptive EPC combines the process of the system identification algorithm and the generalized predictive control design into a

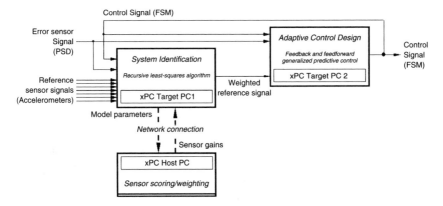

Figure 2.9   Control algorithm and sensor design as implemented on DSP platforms

single algorithm. The combined optical/structural test bed used for this study was described previously and illustrated in Figure 2.2.

2.4.3.1   Digital Signal Processing (DSP) Setup

The application algorithm loaded on each DSP platform is shown in Figure 2.9. MATLAB was used for control system design and implementation in conjunction with an xPC target. Note that the complex nature of this effort and hardware limitations necessitated the use of two target computers. The RLS system identification algorithm was loaded on the xPC target computer and executed in real time. The xPC host computer is connected to the xPC target computer through the network and performs the reference sensor optimization process. Due to the time delay in the data transfer between the target PC and the host PC as well as the computational complexity of the optimization process, the optimal pair of sensors is fixed until a new set is computed and updated. This updating occurs at a slower rate than the system identification process. The weighted reference sensor signal is then computed from the optimal sensor pair and sent to the adaptive controller implemented on a second xPC target computer at a consistent sample rate.

2.4.3.2   Experimental Results

A sketch of the experimental test bed is depicted in Figure 2.10, providing details of accelerometer, actuator, and mounting post coordinates on the flexible plate illustrated previously in Figure 2.2. Prior to the application

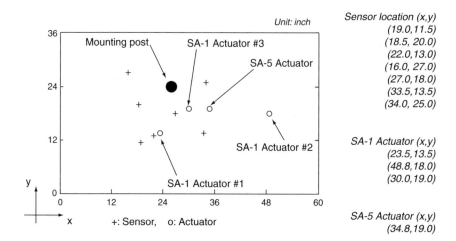

Figure 2.10 Top-view sketch of accelerometers, actuators and mounting post on the plate

of the proposed control algorithm, a band-limited (0–2000 Hz) stochastic disturbance input was applied to actuator #1 mounted on the plate as illustrated in Figure 2.10, and the vibration response was measured from seven accelerometer sensors using a Siglab spectrum analyzer. Figure 2.11(a) shows the averaged magnitude of the frequency response for all accelerometers and Figure 2.11(b) shows the averaged magnitude of the coherence estimation between the disturbance signal and accelerometer signals. As seen in Figure 2.11(b), coherence near 0 dB is measured above 300 Hz. With this observation, the selection vector for the targeted modes in the bandwidth, Λ_c in Equation (2.8), is chosen such that it includes all the modes between 300 and 800 Hz of an identified model. The selection vector for the unimportant modal coupling, Λ_d, in Equation (2.9), is chosen to include all the modes below 300 Hz.

2.4.3.3 Ordered Reference Sensor Selection

First, the adaptive control system using Method 1 as illustrated in Figure 2.8(a) was applied to the test bed. A single actuator (actuator #1 in Figure 2.10) was used as a disturbance input to the plate with a band-limited (0–2000 Hz) stochastic input. The fast steering mirror was used as the control input, and the PSD was implemented in the feedback path of the GPC. The accelerometer ranking first in the scoring metric was incorporated as the reference sensor in the feedforward path of the GPC algorithm. Figure 2.12(a) shows the magnitude of the averaged open- and closed-loop

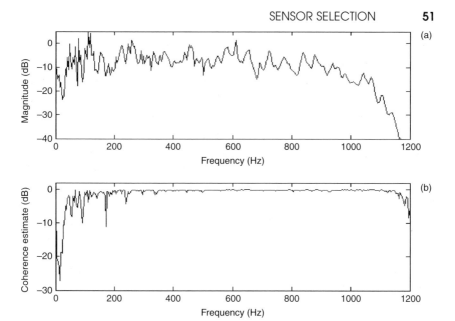

Figure 2.11 (a) Frequency response plot of the averaged magnitude and (b) averaged coherence estimation between the disturbance signal and accelerometer signals

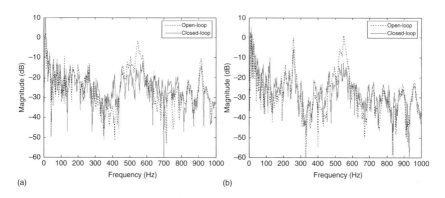

Figure 2.12 Magnitude of frequency response between disturbance input and PSD output for application of the high-ranked reference sensor signal when (a) actuator #1 is disturbance and (b) actuator #2 is disturbance

frequency response between the disturbance source and the PSD signal. Although spillover is observed in some frequency ranges resulting from implementation of the control algorithm, the overall response is attenuated by 6.6 dB between 300 and 800 Hz.

While the control algorithm is running in real time, the disturbance source was switched to an alternative actuator (actuator #2 in Figure 2.10). During this process, both the accelerometer reference sensor employed in the feed-forward control path and the dynamic compensator change. Figure 2.12(b) shows the magnitude of the averaged open- and closed-loop frequency response between the disturbance source and the PSD signal when the disturbance source is modified. The response is attenuated by 9.6 dB between 300 and 800 Hz.

In order to see the performance of the proposed ordered reference sensor selection algorithm (Method 1), the low-ranked accelerometer signal was incorporated instead as the reference sensor in the feedforward path of the GPC algorithm with a consistent level of control effort. The overall response is attenuated by only 2.2 dB between 300 and 800 Hz. Figure 2.13 shows the frequency response between the disturbance source and the PSD signal when actuator #2 is used as the disturbance source actuator. Comparing Figure 2.13 to Figure 2.13(b), it is clear that significant improvement is achieved by using the high-ranked accelerometer sensor from the scoring metric in the feedforward control path.

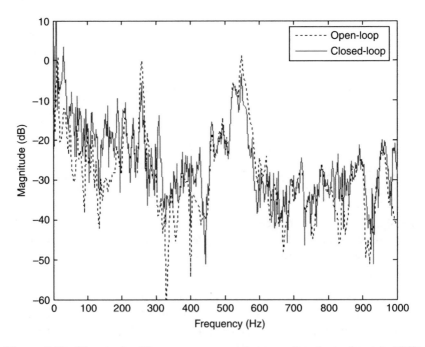

Figure 2.13 Magnitude of frequency response between disturbance input and PSD output for application of the low-ranked reference sensor signal when disturbance source is actuator #1

2.4.3.4 Weighted Reference Sensor Selection

In the second experimental configuration, the adaptive control system using the weighted reference sensor of Method 2 (illustrated in Figure 2.8(b)) was applied to the test bed. The plate structure is disturbed by a band-limited (0–2000 Hz) stochastic disturbance. The disturbance signal is applied to one of the actuators mounted on the flexible plate (actuator #1 in Figure 2.10). Figure 2.14(a) shows the magnitude of the averaged open- and closed-loop frequency response between the disturbance source (actuator #1) and PSD signal. The overall response is attenuated by 5.3 dB between 300 and 800 Hz, and by 6.3 dB between 400 and 600 Hz.

While the algorithm is running in real time, the disturbance source is switched to another actuator (actuator #3 in Figure 2.2). Actuator #3 is also excited with a band-limited (0–2000 Hz) stochastic disturbance. Figure 2.14(b) shows the magnitude of the averaged open- and closed-loop frequency response between actuator #3 and the PSD signal. The overall response is attenuated by 6.2 dB between 300 and 800 Hz, and by 6.7 dB between 400 and 600 Hz.

In order to demonstrate the performance of the proposed weighted reference sensor selection algorithm (Method 2), the closed-loop response was obtained when actuator #1 and the SA-5 actuator (see Figure 2.2) were used as disturbance sources. Uncorrelated band-limited stochastic input signals were applied to each actuator. The averaged open-loop auto spectral density of the PSD signal is represented by the dotted line in Figure 2.15. The averaged closed-loop auto spectral density of the PSD signal is illustrated

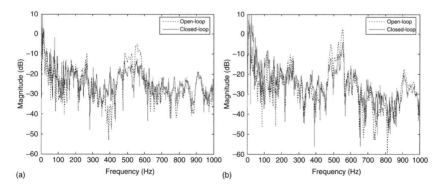

Figure 2.14 Magnitude of frequency response between distrubance input and PSD output for application of the weighted reference sensor signal when (a) actuator #1 is distrubance and (b) actuator #3 is disturbance

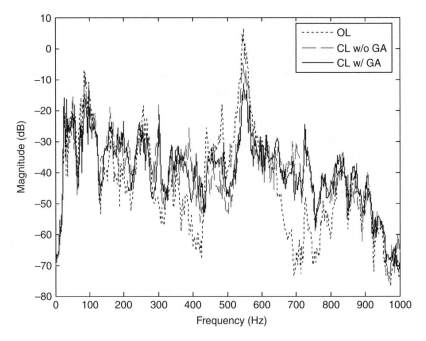

Figure 2.15 Auto spectral density of PSD output for application of the weighted reference sensor signal: open-loop (dotted line), closed-loop with single fixed reference sensor (dashed line), and closed-loop with the weighted reference sensor signal (solid line)

in Figure 2.15 by the solid line for the weighted sensor array. The overall response is attenuated by 8.8 dB between 300 and 800 Hz. For comparison, the sensor with the highest score in the rank-ordered sensor set was implemented as a reference signal in the feedforward path of the GPC algorithm. The averaged closed-loop auto spectral density of the PSD signal is shown as the dashed line in Figure 2.15 for an equivalent control effort. The overall response is attenuated by 4.3 dB between 300 and 800 Hz. As illustrated in Figure 2.15, for the experimental example presented, the weighted reference sensor selection algorithm achieves better performance than the single reference sensor for the same level of control effort.

2.4.3.5 Summary

Two alternative selection/weighting methods for sensor arrays have been presented. In both cases, the design metric was based upon estimated Hankel singular values as a mechanism for identifying sensors that best reflect the dynamics/modal characteristics of the error sensor. The design/selection was implemented in real time, in parallel with an adaptive controller to demonstrate the potential of simultaneous temporal and spatial compensation. The implementation of such requires that the time-varying dynamics of the system or changes in disturbance path are slower than that of the sample rate of the controller. In all cases presented, both the rank-ordered sensors and the weighted sensors yielded greater performance for comparable control effort than non-optimized sensors. Furthermore, real-time adaptation of the sensors was demonstrated.

2.5 CONCLUSIONS

In distributed sensing for active control, knowledge of the number of independent disturbance sources can effectively guide the control system designer's approach to feedback and feedforward control. In an effort to reduce the number of sensors incorporated in a control system without compromising performance, one must determine the location and minimal number of sensors required to capture the full bandwidth response of the disturbance(s) through either first-principle models or experimentally identified models. The first part of this chapter was devoted to the application of principal component analysis and the identification of the number of disturbance sources. As was demonstrated, the magnitude of the disturbance as measured at the sensors can greatly affect the results obtained from principal component analysis, and active control can be used to identify additional disturbance sources, particularly if the sensor response due to these disturbances is an order of magnitude or more below that of the primary disturbance.

Once an appropriate array of sensors capable of capturing the plant response to exogenous disturbances has been identified, a mechanism for optimizing this array so as to minimize the number of inputs to the control system is desired. Furthermore, finding an optimal subset or a weighted output of the array that conveys the essential dynamics sought in the performance path can greatly enhance performance while reducing sensor count and simplifying control system design. It is possible through design optimization to further select/design sensors that couple only to targeted modes represented in the performance metric and ignore modes unimportant for control. Results from the experimental implementation of a real-time sensor selection/weighting algorithm demonstrate that it is feasible to adapt the temporal compensation and spatial compensation in parallel to enhance performance for varying disturbance inputs. Basic design procedures were outlined for simultaneous spatial and temporal compensation.

ACKNOWLEDGMENTS

The authors would like to thank the Air Force Research Laboratory at Kirtland Air Force Base in Albuquerque, NM and the Missile Defense Agency, contract No. FA945304C0013, for providing funding for this research.

REFERENCES

[1] J. Luo and Z. Zhang. 'Using Eigenvalue Grads Method to Estimate the Number of Signal Source', Proceedings of ICSP2000, pp. 223–225, 2000.

[2] A. Paulraj and T. Kailath 'Eigenstructure Methods for Direction of Arrival Estimation in the Presence of Unknown Noise Fields', IEEE Transactions on Acoustics, Speech, and Signal Processing, **ASSP-34**(1), February, pp. 13–20, 1986.

[3] T.-J. Shan, M. Wax, and T. Kailath. 'On Spatial Smoothing for Direction-of-Arrival Estimation of Coherent Signals', IEEE Transactions on Acoustics, Speech, and Signal Processing, **ASSP-33**(4), August, pp. 806–811, 1985.

[4] H. Akaike. 'Information Theory and an Extension of the Maximum Likelihood Principle', in Proceedings 2nd International Symposium Information Theory, suppl. Problems of Control and Information Theory, 1973, pp. 267–281.

[5] J. Rissanen. 'Modeling by Shortest Data Description', Automatica, **14**, pp. 465–471, 1978.

[6] S.-M. Moon, R. L. Clark, and D. G. Cole. 'The Recursive Generalized Predictive Feedback Control: Theory and Experiments', Journal of Sound and Vibration, **279**, pp. 171–199, 2005.

[7] L. Ljung. *System Identification: Theory for the User*, Prentice Hall, Englewood Cliffs, NJ, 2nd edition, 1999.

[8] K. B. Lim, R. C. Lake, and J. Heeg. 'Effective Selection of Piezoceramic Actuators for an Experimental Flexible Wing', AIAA Journal of Guidance, Control and Dynamics, **21**(5), pp. 704–709, 1998.

[9] R. E. Richard and R. L. Clark. 'Computationally Efficient Piezostructure Modeling for System Optimization', Proceedings of SPIE: Smart Structures and Materials, **4701**, pp. 389–400, July 2002.

[10] S.-M. Moon, D. G. Cole, and R. L. Clark. 'Real-time Implementation of Adaptive Feedback and Feedforward Generalized Predictive Control Algorithm, Journal of Sound and Vibration, **294**, pp. 82–96, 2006.

[11] R. L. Clark, G. P. Gibbs, and W. S. Saunders. *Adaptive Structures: Dynamics and Control*, John Wiley & Sons, Inc. New York, 1998.

[12] R. E. Richard and R. L. Clark. 'Delta Wing Flutter Control Using Spatially Optimized Transducers', Journal of Intelligent Material Systems and Structures, **11**(14), pp. 677–691, 2003.

[13] K. B. Lim and W. Gawronski. 'Hankel Singular Values of Flexible Structures in Discrete Time', AIAA-1996-3757, Guidance, Navigation and Control Conference, San Diego, CA, July 29–31, 1996.

[14] J. N. Juang, M. Phan, L. G. Horta, and R. W. Longman. 'Identification of Observer/Kalman Filter Markov Parameters – Theory and Experiments', Journal of Guidance, Control, and Dynamics, **16**(2), pp. 320–329, 1993.

[15] K. B. Lim and W. Gawronski. 'Hankel Singular Values of Flexible Structures in Discrete Time', ASME Journal of Guidance, Control, and Dynamics, **19**(6), pp. 131–145, 1996.

[16] D. E. Goldberg. *Genetic Algorithms in Search, Optimization, and Machine Learning*, Addison-Wesley, Reading, MA, 1989.

[17] D. W. Clarke, C. Mohtad, and P. S. Tuffs. 'Generalized Predictive Control—Part I. The Basic Algorithm', Automatica, **23**(2), pp. 137–148, 1987.

[18] D. W. Clarke, C. Mohtad, and P. S. Tuffs. 'Generalized Predictive Control—Part II. Extensions and Interpretations', Automatica, **23**(2), pp. 149–160, 1987.

[19] W. Gawronski, and T. Williams. 'Model Reduction for Flexible Space Structures', Journal of Guidance, Control, and Dynamics, **14**(1), pp. 68–76, 1991.

[20] R. L. Clark and D. E. Cox. 'Experimental Demonstration of a Band-Limited Actuator/Sensor Selection Strategy for Structural Acoustic Control', Journal of the Acoustical Society of America, **106**(6), pp. 3407–3414, 1999.

[21] R. L. Clark and D. E. Cox. 'Band-Limited Actuator and Sensor Selection for Disturbance Rejection: Application to Structural Acoustic Control', AIAA Journal of Guidance, Control, and Dynamics, **22**(5), pp. 740–743, 1999.

[22] S. M. Moon, L. P. Fowler, and R. L. Clark. 'Optimal Sensing Strategy for Adaptive Control of Optical Systems', Proceedings of SPIE: Smart Structures and Materials 2005: Modeling, Signal Processing, and Control, Ralph C. Smith, Editor, **5757**, pp. 209–218, 2005.

3

Global Vibration Control Through Local Feedback

Stephen J. Elliott

Institute of Sound and Vibration Research, University of Southampton, Southampton SO17 1BJ, UK

3.1 INTRODUCTION

The control of vibration in systems with many degrees of freedom requires multiple actuators and multiple sensors. Conventionally, all the actuators are driven by a single, centralised controller, which is also supplied with signals from all the sensors, as shown on the left hand side of Figure 3.1. Such centralised controllers are generally designed using a model of the system under control, such as a modal model, and have the advantage that the controller can be designed so that individual modes can be influenced to different extents (Meirovitch, 1990; Fuller *et al.*, 1996; Clark *et al.*, 1998; Preumont, 2002). The type of actuators and sensors used and their positioning are chosen to best influence or observe these modes. Such controllers can perform very well if the dynamics of the system under control are of relatively low order and are well known, as are the disturbances acting on the system. Potential disadvantages of centralised controllers are that, first, their performance, and sometimes their stability, can be threatened if the response of the system under control changes, or if a transducer fails, so that the assumed model is no longer an accurate one, and, second, a great deal of rapid communication is required in large systems to connect all the actuators

Adaptive Structures: Engineering Applications   Edited by D. Wagg, I. Bond, P. Weaver and M. Friswell
© 2007 John Wiley & Sons, Ltd

and sensors to the single controller, and the complexity of the controller rises sharply as the size of the system increases. Decentralised controllers are thus particularly worth considering for the control of vibrations at frequencies where many modes are contributing, such as at audio frequencies, on structures whose dynamics are uncertain and potentially time varying, subject to unknown disturbances.

In a fully decentralised system a number of independent controllers are used to drive individual actuators from individual sensors, as shown on the right hand side of Figure 3.1. Care must be taken in selecting the type of actuator and sensor pair used and their close proximity to ensure stability of the multiple independent loops when working simultaneously. Such a control strategy has the advantages of reduced complexity and wiring and, if all the controllers are the same, it also has the advantage of modularity, so that increasingly complex systems can be controlled with a number of identical simple units, and scalability, so the complexity of the system scales only linearly with the number of modes. Because such units only act locally, however, individual global properties of the system, such as mode amplitudes, cannot be selectively controlled.

In Sections 3.2 and 3.3 of this chapter the centralised and decentralised control of vibration on a structure will be compared, and the attractive stability properties of local feedback control with dual and collocated pairs of actuators and sensors will be discussed. It is shown that the performance of a decentralised controller with such transducers in reducing the global response

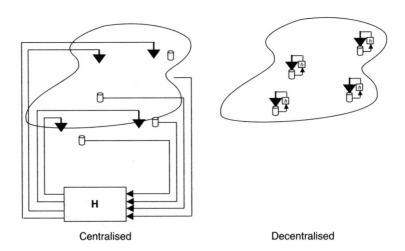

Centralised   Decentralised

Figure 3.1 Centralised and decentralised control systems for the active control of vibration with four actuators (arrows) and four sensors (cylinders)

of the system can be almost as good as a centralised controller connecting these transducers. This is illustrated in Section 3.4 for the control of vibration on a plate with distributed excitation. The tuning of the local controller gains is also discussed, together with the stability difficulties associated with using practical piezoelectric or inertial actuators, particularly with large numbers of loops.

Whereas negative feedback is used to reduce the vibration in engineering structures, it is becoming clear that positive feedback is used by nature to enhance the vibration of the basilar membrane in the inner ear, which improves the sensitivity and selectivity of our hearing, as discussed in Section 3.5. These feedback loops are heavily nonlinear, which has a profound effect on our perception of sound. The enhancement relies on a large number of locally acting feedback loops being tuned close to instability and their gains being regulated in ways that are not well understood.

3.2   CENTRALISED CONTROL OF VIBRATION

The conventional approach to controlling the dynamics of large structures, with many degrees of freedom, is to work in terms of the structure's modes (Meirovitch, 1990; Gawronski, 1998). The dynamic behaviour is approximated by the sum of a finite number of modal contributions, actuators and sensors are chosen so that they can observe and control these modes, and a feedback controller is designed to modify their response. Often the objective is not actively to change the shapes of the modes, but to modify independently the natural frequencies and damping ratios of a number of modes of interest, which has been termed independent modal-space control, IMSC, by Meirovitch (1990).

A schematic diagram of a multichannel feedback controller for modal control is shown in Figure 3.2. The sensor outputs are processed by a modal analyser to provide estimates of the instantaneous amplitudes of the controlled modes and these are fed via a diagonal modal controller to a modal synthesiser, which drives each of the actuators to excite the structure with the required modal amplitudes. The great advantage of such an architecture is that the design of the controller is relatively straightforward, since it is diagonal and, generally, only displacement and velocity in each mode are fed back to modify the modal stiffness and damping. The modal analyser and synthesiser are designed from a knowledge of the dynamics of the structure and are generally fully populated matrices with constant real coefficients. The modal approach thus generates a fully coupled feedback controller for a

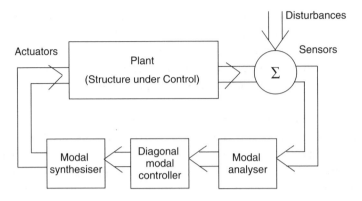

Figure 3.2 Block diagram of a modal feedback controller used to control the vibration of a structure

multiple input, multiple output (MIMO) system, which is relatively easy to design and whose action has a clear physical interpretation.

In practice the structure under control will not respond only in the modes assumed in the design of the controller, however, since an infinite number of modes are required to describe exactly the response of a distributed system. The modelling of structural dynamics often relies on this infinite modal sum being reasonably well approximated by a finite sum and so modal control fits in well with conventional dynamic analysis. The effect of the 'residual' modes that are not taken into account in the design of the modal controller can be accounted for in terms of observation and control spillover, whose effects are reasonably well understood (Meirovitch, 1990). Using a decomposition in terms of the controlled and the uncontrolled modes, the MIMO "plant" response, $\mathbf{G}(s)$, in Figure 3.2, from all the actuators to all the sensors on the structure under control, can be written as

$$\mathbf{G}(s) = \Psi \, \Lambda(s) \, \Phi + \mathbf{R}(s) \tag{3.1}$$

where Ψ is the matrix of controlled mode shapes at the sensor positions, Φ is the matrix of controlled mode shapes at the actuator positions, which are assumed to be real, $\Lambda(s)$ is the diagonal matrix of controlled mode amplitudes and $\mathbf{R}(s)$ is a matrix that accounts for the influence of the uncontrolled modes.

The overall response of the modal feedback controller in Figure 3.2, $\mathbf{H}(s)$, can be written as

$$\mathbf{H}(s) = \Phi^{\dagger} \, \Omega(s) \, \Psi^{\dagger}, \tag{3.2}$$

where $\boldsymbol{\Psi}^\dagger$ and $\boldsymbol{\Phi}^\dagger$ are the responses of the modal analyser and modal synthesiser, which may be the pseudo-inverses of $\boldsymbol{\Phi}$ and $\boldsymbol{\Psi}$ in Equation (3.1) (Fuller *et al.*, 1996), and $\boldsymbol{\Omega}(s)$ is the diagonal matrix of modal control terms. The modal analyser and synthesiser depend only on the shapes of the structure's modes and these are known to vary less under perturbation than the natural frequencies. Provided the mode shapes have been correctly identified in the first place, the basic structure of the modal controller shown in Figure 3.2 is thus still valid even if the structure is perturbed to some extent.

In a well-defined structure it may not be difficult to model numerically the first 10 or so modes, using finite elements for example, or to identify experimentally a similar number of modes using modal analysis. It becomes increasingly difficult, however, to distinguish between the different modal contributions beyond this number, particularly in built-up structures with a high modal overlap. Under these conditions the modal approach to control becomes inappropriate. If we are considering the control of large structures, particularly at the higher frequencies for which sound radiation is important, then a reliable modal model of the structure is generally not available and alternative approaches to active control have to be found.

3.3 *DECENTRALISED CONTROL OF VIBRATION*

The modal controller described in the previous section is a particular form of centralised controller, in which every actuator must be connected to every sensor via a single control unit. The disadvantages of such a system in terms of wiring and robustness to failure were described in the Introduction. A fully decentralised vibration controller would require that each actuator was only driven by the output of a single sensor and that suitable pairs of actuators and sensors had been identified for such local controllers. In fact decentralised control has some very attractive properties if the actuator/sensor pairs are collocated and so the choice of actuator and sensor positions are not independent, and the pairing problem (Friswell *et al.*, 1996) disappears. We will additionally assume, for the time being, that the actuators and sensors are perfect and are also dual (Sun, 1996): that is, the product of the sensor output and actuator input, if used in isolation, is proportional to the power supplied to the system. We initially assume collocated force actuators and velocity sensors as a particular example of dual actuator/sensor pairs.

If local, fixed-gain, control loops are closed around such actuator/sensor pairs, the control system is an example of direct output feedback control and has some very attractive stability properties (Balas, 1979). Since each local feedback loop can only absorb power from the structure, it is an example of a dissipative controller (Benhabib *et al.*, 1981; Gawronski, 1998), and no combinations of positive gains can destabilise the system. Each force actuator driven by a collocated velocity sensor synthesises an entirely passive damper. In practice vibration velocity is often estimated by integrating the output of an accelerometer, and so the velocity is measured relative to the inertial reference frame. The integrator singularity at dc is avoided since rigid-body modes are not being controlled. The actively synthesised dampers thus operate with respect to this reference frame and are called skyhook or inertial dampers (Karnopp, 1995).

The robustness of such a decentralised system can be illustrated by considering the behaviour of a single channel system in the frequency domain, so that $G(j\omega)$ is the plant response from secondary force input to velocity sensor output. In this case, $G(j\omega)$ is equal to the input, or point, mobility of the structure, so that $v = G(j\omega)f$ where v is the structure's velocity and f the applied force. The mechanical power supplied to the structure by this force is

$$W = \tfrac{1}{2}\operatorname{Re}[f^*v] = \tfrac{1}{2}|f|^2 \operatorname{Re}[G(j\omega)] \qquad (3.3)$$

and since W must be positive then so must $\operatorname{Re}[G(j\omega)]$, and the plant is said to be passive. The stability properties of the feedback loop can be described in terms of the system's Nyquist plot. Since the feedback controller, $H(j\omega)$, is a positive constant, h, for direct velocity control, and the real part of $G(j\omega)$ is also positive, the Nyquist plot lies entirely on the right hand side of the imaginary axis and is thus well away from the instability point $(-1, 0)$, so that the control strategy is inherently robust.

One has to be more careful in analysing the stability of a multichannel decentralised feedback system. The feedback loops can no longer be regarded as independent since they are coupled via the overall dynamics of the structure. The stability of a multichannel feedback control system can be determined using a generalisation of the Nyquist plot, in which the locus of the real and the imaginary parts of eigenvalues of the matrix $\mathbf{G}(j\omega)\mathbf{H}(j\omega)$ are plotted as ω varies from $-\infty$ to ∞ (Skogestad and Postlethwaite, 1996), where $\mathbf{G}(j\omega)$ and $\mathbf{H}(j\omega)$ are the matrices of plant responses and feedback controller responses. In this case the controller is decentralised and each loop is assumed to have a constant gain, so that $\mathbf{H}(j\omega)$ is equal to $h\mathbf{I}$. The vector of velocities, $\mathbf{v}$, is related to the vector of actuator forces, $\mathbf{f}$, by $\mathbf{v} = \mathbf{G}(j\omega)\,\mathbf{f}$, where $\mathbf{G}(j\omega)$ is the fully populated plant response. If the actuators and sensors are collocated, $\mathbf{G}(j\omega)$ is the input mobility matrix for the structure, which is symmetric due to reciprocity.

The total mechanical power supplied by all the actuator forces can then be written as

$$W = \tfrac{1}{2} \operatorname{Re}\left[\mathbf{f}^H \mathbf{v}\right] = \tfrac{1}{2} \operatorname{Re}\left[\mathbf{f}^H \mathbf{G}(j\omega)\,\mathbf{f}\right] \tag{3.4}$$

and since W must be positive, then $\mathbf{G}(j\omega)$ must be passive. In this case passivity implies that both

$$\operatorname{Eig}\left[\operatorname{Re}\,\mathbf{G}(j\omega)\right] \geq 0 \tag{3.5a}$$

and

$$\operatorname{Re}\left[\operatorname{Eig}\,\mathbf{G}(j\omega)\right] \geq 0. \tag{3.5b}$$

The second condition is less well known than the first and can be demonstrated by writing the eigendecomposition of the normal, symmetric, but complex matrix $\mathbf{G}$ as $\mathbf{Q}^H \boldsymbol{\Lambda} \mathbf{Q}$ where $\mathbf{Q}$ and $\boldsymbol{\Lambda}$ are generally complex. The power supplied to the system is thus proportional to

$$\operatorname{Re}\left[\mathbf{f}^H \mathbf{G}\,\mathbf{f}\right] = \operatorname{Re}\left[\mathbf{q}^H \boldsymbol{\Lambda}\,\mathbf{q}\right] = \operatorname{Re}\left[\sum_i \lambda_i\,|q_i|^2\right] \tag{3.6}$$

where $\mathbf{q} = \mathbf{Q}\mathbf{f}$, and since the power must be positive for all inputs, so must the real parts of the eigenvalues of $\mathbf{G}$. This condition ensures that each of the loops in the generalised Nyquist diagram behave in a similar way to those of the single channel system and are strictly on the right hand side of the imaginary axis, ensuring that the control system is very robust. In particular, the stability is ensured for any degree of flexibility in the structure, no matter how strongly the actuators and sensors are coupled. Exactly the same argument can be used to show that a control system with one fewer control loops is robustly stable, and so the stability of the system is not compromised by failure of individual control loops.

Ideal local velocity control thus has very attractive robustness properties, since it is unconditionally stable, whatever the response of the structure to which it is attached, and in spite of any failures that may occur in other control loops. The design of such local loops to achieve a specific performance objective in controlling the structure is less obvious than it was for modal control, however. Clearly the actuator/sensor pairs should be positioned so that they efficiently couple into the modes to be controlled. The individual gains could be optimised to minimise a specific quadratic cost function, using state-space methods for example. In general, however, decentralised control is not selective, since it does not allow the independent specification of the modal responses that is the attractive feature of modal control. If, for example, a large number of force actuator/velocity sensor pairs with equal feedback gains are uniformly distributed over a structure then the effect is to increase the damping of an equal number of the structural modes to approximately the same extent.

In order to justify this statement we initially return to the centralised modal controllers shown in Figure 3.2. In the special case in which there are an equal number of collocated force actuators and velocity sensors, then $\mathbf{G}$ and $\mathbf{H}$ will be square matrices and Ψ will be equal to Φ^T, so that, assuming the effect of the uncontrolled modes is small, then

$$\mathbf{G}(s) \approx \Phi^T \Lambda(s) \Phi. \tag{3.7}$$

We also assume that there are the same number of controlled modes as sensors or actuators and that Φ is non-singular, so that the modal controller in Equation (3.2) becomes

$$\mathbf{H}(s) = \Phi^{-1} \Omega(s) \Phi^{-T}. \tag{3.8}$$

If the proportional components in each diagonal term of $\Omega(s)$ are positive, the controller is passive, and since the plant is also passive, under these conditions the system is unconditionally stable.

Assuming the system is stable, the vector of output signals for any multi-channel controller can be written as

$$\mathbf{y}(j\omega) = [\mathbf{I} + \mathbf{G}(j\omega)\mathbf{H}(j\omega)]^{-1}\mathbf{d}(j\omega) \tag{3.9}$$

where $\mathbf{d}(j\omega)$ is the vector disturbance.

The vector of output signals for the centralised modal controller can thus be written as

$$\mathbf{y}(j\omega) \approx \left[\mathbf{I} + \Phi^T \Lambda(j\omega) \Omega(j\omega) \Phi^{-T}\right]^{-1} \mathbf{d}(j\omega) \tag{3.10}$$

so that

$$\Phi^{-T}\mathbf{y}(j\omega) \approx [\mathbf{I} + \Lambda(j\omega) \Omega(j\omega)]^{-1} \Phi^{-T}\mathbf{d}(j\omega), \tag{3.11}$$

where the elements of $\Phi^{-T}\mathbf{y}(j\omega)$ now represent the modal amplitudes of the system. It is clear that the response of each of the controlled modes can be individually controlled by a suitable choice of terms in each of the diagonal elements of $\Omega(j\omega)$.

The effect of decentralised control with collocated force actuators and velocity sensors on a structure with an equal number of controlled modes, so that Equation (3.7) is still valid, can be established by assuming that $\mathbf{H} = h\,\mathbf{I}$. The system is stable if $h \geq 0$ and the closed loop response can be written as

$$\mathbf{y}(j\omega) \approx \left[\mathbf{I} + h\,\boldsymbol{\Phi}^{\mathrm{T}}\,\boldsymbol{\Lambda}(j\omega)\,\boldsymbol{\Phi}\right]^{-1}\mathbf{d}(j\omega). \tag{3.12}$$

We now assume that there are a large number of actuators and sensors, which are reasonably uniformly distributed, in which case the matrix $\boldsymbol{\Phi}\,\boldsymbol{\Phi}^{\mathrm{T}}$ tends to the identity matrix because of the orthogonality of the modes (Bullmore et al., 1987), so that $\boldsymbol{\Phi} \approx \boldsymbol{\Phi}^{-\mathrm{T}}$, again assuming that $\boldsymbol{\Phi}$ is not singular. With this approximation, the vector of mode amplitudes can be written as

$$\boldsymbol{\Phi}^{-\mathrm{T}}\mathbf{y}(j\omega) \approx \left[\mathbf{I} + h\,\boldsymbol{\Lambda}(j\omega)\right]^{-1}\boldsymbol{\Phi}^{-\mathrm{T}}\mathbf{d}(j\omega). \tag{3.13}$$

By comparing this equation with Equation (3.11) it is clear that under these ideal circumstances the decentralised controller has exactly the same effect as a full modal controller with equal constant feedback gains, i.e. $\boldsymbol{\Omega}(j\omega) = h\,\mathbf{I}$.

3.4 CONTROL OF VIBRATION ON STRUCTURES WITH DISTRIBUTED EXCITATION

Local velocity feedback has been successfully used to control vibration transmission in active isolation systems (Serrand and Elliott, 2000; Kim et al., 2001, Huang et al., 2003). In this case, illustrated in Figure 3.3, the velocity sensors are placed immediately above the mounts, which are the only path of vibration transmission from the base to the receiving structure. For the best isolation these velocities should be attenuated to the greatest possible extent and so the feedback gains in each of the decentralised loops should be as large as possible.

In this section we examine the use of local control to reduce the vibration of structures with a distributed excitation, so that the forcing cannot be counteracted at source. An example would be a plate excited by a random pressure field, as illustrated in Figure 3.4. There are several feedback control strategies that could be used to connect the multiple sensors to the multiple actuators for such active vibration control. An interesting study was conducted by Petitjean and Legrain (1996), in which an array of 3×5 piezoelectric patches on one side of a plate were used as actuators and an identical array of patches were used as sensors on the other side. Petitjean and Legrain used a state-space model of the plate to design a fully connected modal controller for

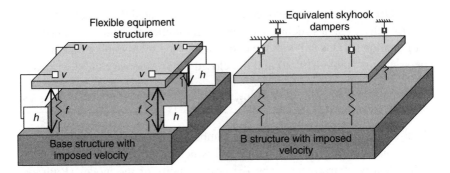

Figure 3.3   A multichannel active skyhook damper for the isolation of base vibration, in which the velocity at each mount, indicated as an ideal spring, is fed back only to a force actuator in parallel with the passive mount (left); the equivalent mechanical system (right)

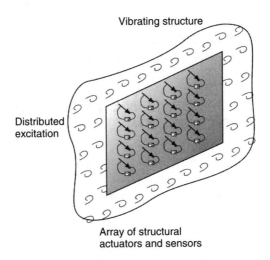

Figure 3.4   Active control of a plate with a distributed excitation using an array of 16 decentralised feedback controllers driving secondary force actuators (arrows) from collocated velocity sensors (circles)

the system. They also tested the performance of a completely decentralised controller, in which each actuator was only driven by the sensor on the opposite side of the plate, and found that the practical performance of the decentralised system was very similar to the fully centralised system. This observation further motivated the study of active vibration control systems with many independently operating feedback loops. In order to understand the physical action of such systems, Elliott *et al.* (2002) considered the effect of a 4×4 array of collocated force actuators and velocity sensors in the vibration control of a thin aluminium plate (278 mm × 247 mm × 1 mm).

The stability of the control system is guaranteed provided each of the individual gains remains positive, since the plant response is again passive with the assumed collocated force actuators and velocity sensors. In this preliminary study all of the gains in the individual feedback loops were assumed equal. The expected value of the kinetic energy of the plate when excited by a temporally and spatially random pressure field is shown as a function of excitation frequency in Figure 3.5. The solid line shows this without control, and the peak at about 70 Hz is due to resonances of the (1,1) mode of the plate. As the gains of the decentralised control units are all increased, the modal peaks are suppressed, and the overall kinetic energy due to broadband excitation up to 1 kHz is also decreased, as shown in Figure 3.6. Further increasing all the feedback gains reduces the kinetic energy at low frequencies, but for large gains, the kinetic energy at higher frequencies, at about 600 Hz and above in this simulation, starts to increase, as seen in Figure 3.5. At very high gains the action of the individual control loops is

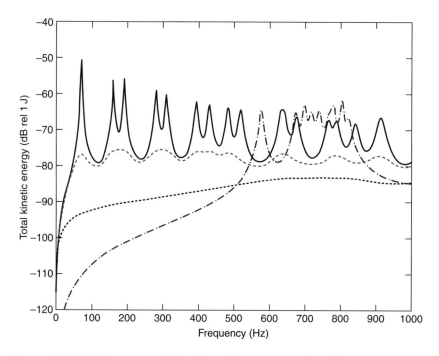

Figure 3.5 Kinetic energy of the plate with no control (solid) and for feedback gains of 10 (dashed), 100 (dotted) and 1000 (dot dash) using 16 feedback loops with force actuators and velocity sensors

to *pin* the plate at the sensor locations, so that the modes of the modified structure occur at higher frequencies, but are not very heavily damped by the action of the feedback controllers.

In this application, the excitation of the vibrating structure is distributed, so the actuator sensor pairs cannot be positioned to block the excitation completely with high-gain controllers. With a finite number of actuator/sensor pairs on a structure with distributed excitation, the feedback gain is thus a trade-off between damping the modes of the original structure and not generating additional lightly damped modes, with the best trade-off being obtained at the optimum gain.

The interesting question then arises as to whether this optimum gain can be deduced from purely local measurements, since in this case each individual feedback loop could tune its own gain, based only on the input to its own actuator and the output of its own sensor. One possible way of achieving this is suggested by the observation that the optimum feedback gain is approximately equal to the impedance of the infinite structure (Elliott *et al.*, 2002), and the additional observation that the global performance is relatively insensitive to the exact value of the gain, being degraded by only 3 dB for a factor of 10 variation in the gain about its optimal value in Figure 3.6.

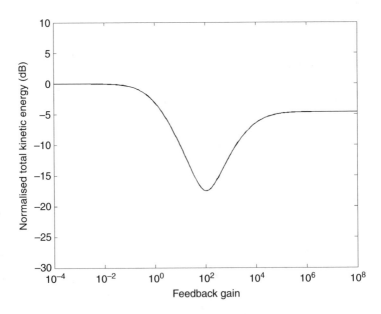

Figure 3.6 Change in the average kinetic energy, integrated from 0 kHz to 1 kHz, as a function of the equal gain in each of the 16 decentralised loops on the plate with force actuators and velocity sensors

Thus, if an initial identification process is performed in an individual loop to determine the open loop response between the output of the velocity sensor and the input to the force actuator, which is equal to the point mobility of the structure, and the mobility of the infinite system could be deduced from this, then the real gain could be set from this measurement. The point mechanical impedance of an infinite plate is real and frequency independent (Cremer and Heckl, 1987). Engles (2006) has used its estimated value, calculated from the frequency average of the real part of the point impedance, to self-tune a local feedback system on a plate with some success. Another strategy for tuning the feedback gain of a single control loop is based on a trade-off between attenuating one mode and enhancing another (Variyart *et al.*, 2002). This has been shown to work well on a beam, but is less successful on a plate (Engles, 2006).

Another method of estimating the input impedance of the infinite system, Z_∞, is prompted by the observation that in general the real part of the frequency-average point mobility, $\overline{Y}$, can be written (Cremer and Heckl, 1989) as

$$\mathrm{Re}[\overline{Y}] = \mathrm{Re}\left[\frac{1}{Z_\infty}\right] = \frac{\pi}{2M}\frac{\Delta N}{\Delta \omega}, \qquad (3.14)$$

where M is the total mass of the structure and $\Delta N/\Delta \omega$ is the modal density. For an infinite plate, the point mobility is entirely real, and the modal density is independent of frequency and approximately equal to the reciprocal of the first natural frequency, ω_1, so that

$$Z_\infty \approx \frac{2M\omega_1}{\pi}. \qquad (3.15)$$

If the first natural frequency, ω_1, is written as $\sqrt{4K/M}$, where $M/4$ is the modal mass and K is the low-frequency stiffness of the plate, then

$$Z_\infty \approx \frac{8K}{\pi\omega_1}. \qquad (3.16)$$

The low-frequency stiffness of a plate and its first natural frequency can be readily measured from its input mobility and so Z_∞ could be estimated, and used to tune the feedback control loop.

Finally, another physically attractive strategy for self-tuning local vibration controllers should be mentioned, based on maximising the power absorbed by the active control system (Redman-White *et al.*, 1987; Hirami, 1997; Nelson, 1996). Engles (2006) has compared the performance of this strategy, for a single control loop on a plate, with the infinite impedance and modal approach to self-tuning and found that they have very similar levels of performance, no matter where the control loop is placed on the plate. More surprisingly, he also found that the performance of a dynamic feedback controller designed using LQG methods (as explained, for example, by Meirovitch, 1990) with a full state-space model of the plate did not perform significantly better (Engles, 2006). These results extend to multiple control loops and probably reflect the fact that for ideal force actuators and velocity sensors the best mechanism of vibration control is power absorption, which can only be achieved with local direct feedback.

The simulations of Engles *et al.* (2006) show that for an array of 16 collocated force actuator/velocity sensor pairs on a plate, the frequency-averaged performance of a constant gain decentralised controller is almost as good as a constant gain centralised controller, or even a fully connected optimal LQG controller. Figure 3.7 shows the expectation of the kinetic energy on a plate as a function of frequency when excited by a temporally and spatially random pressure field. Also shown is the kinetic energy with optimally adjusted, constant gain, centralised and decentralised controllers, whose performance is almost identical. The kinetic energy with a fully coupled dynamic feedback controller, designed using LQG methods with

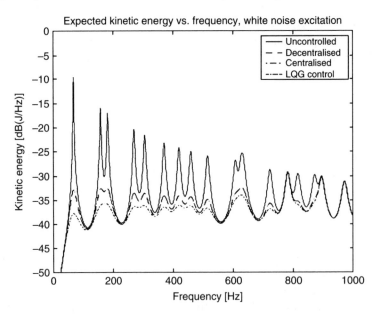

Figure 3.7 Spectrum of the expected kinetic energy of a plate with distributed excitation (solid) and with a centralised or decentralised constant gain controller (dot dash and long dash) or a fully centralised LQG controller (short dash)

a perfect model of the plate dynamics, is also illustrated. Although the performance of this LQG controller is slightly better than that of the constant gain decentralised one at some frequencies, the average performance is very similar. In order to make this comparison fairly, however, the same control effort should be used by the different types of feedback controller, i.e. the same sum of mean squared actuator forces. Figure 3.8 shows the average change in kinetic energy for the different controllers, integrated over the whole frequency range of these simulations, with various limits on the control effort. For the control effort used in Figure 3.7, i.e. $3 \times 10^3 \, \mathrm{N}^2$, the average reduction obtained with all the control strategies was very similar, and this observation is seen to hold over a wide range of values of control effort. This suggests that with these ideal actuators and sensors, local control not only gives good global performance, but also gives a global performance that is almost as good as that of any centralised controller.

In practice it is difficult to generate an idealised force actuation of the type shown in Figure 3.4 without a solid structure nearby to react off. One type of actuator that could be readily integrated into the plate is a piezoceramic patch. The frequency response from a small piezoceramic actuator to an adjacent

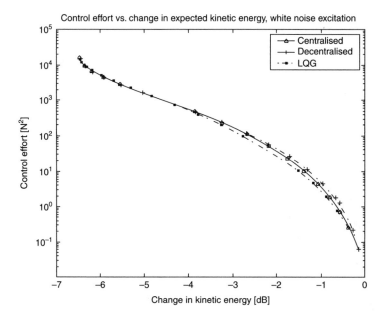

Figure 3.8   The frequency-averaged change in kinetic energy of a randomly excited panel for a centralised (solid) or decentralised (long dash) constant gain controller and a centralised LQG controller (dot dash) for various limits on the control effort, i.e. sum of mean square actuator forces

velocity sensor is surprisingly similar to that of a force actuator to the velocity sensor at low frequencies, and shares the passive property of the latter, with the phase response remaining within ±90°. At higher frequencies, however, the piezopatch actuator is more effective at exciting the plate than the force actuator, and the frequency response to the adjacent velocity sensor does not fall off with frequencies in the same way as it does with the force actuator, making it more difficult to design a stable feedback controller. A practical arrangement of this type has been studied by Gardonio *et al.* (2004) using the integrated output of accelerometers as velocity sensors. They showed that the frequency response, from an individual actuator to an adjacent sensor, is strongly influenced at higher frequencies, where the stability is most threatened, by the size of the piezoceramic actuator, its weight and stiffness, and the added mass and resonance of the accelerometer. Detailed design of the actuator and sensor are thus required to ensure a good performance. By carefully designing the response of the 16 individual feedback loops in such an experimental arrangement, Gardonio *et al.* demonstrated overall reductions, up to 1 kHz, of 9.3 dB in plate velocity, measured with a scanning laser vibrometer, and about 3 dB reduction in radiated sound power, measured with a far-field microphone array in an anechoic chamber.

Another actuator, which approximates an ideal force actuator at high frequencies, is the electromagnetic inertial, or proof-mass, actuator. In this device, illustrated in Figure 3.9, a coil is attached to the case and a magnet provides the magnetic field, but also acts as the inertial mass. The theoretical frequency response of such an inertial actuator, from input voltage to blocked force, is shown in Figure 3.9, and the response of practical devices can follow this ideal response quite closely. At low frequencies the device is stiffness controlled with a response rising as ω^2 with a 180° phase shift,

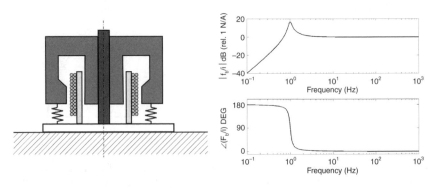

Figure 3.9 Idealised construction of an electromagnetic inertial actuator and its blocked frequency response

until the peak at the resonance is reached, caused by the inertial mass and the suspension stiffness. Above this actuator resonance frequency, the frequency response is flat, with little phase shift.

The phase shift associated with the actuator resonance can destabilise a direct velocity feedback system unless the actuator resonance frequency is well below the first resonance frequency of the structure to which it is attached. This is illustrated in Figure 3.10, in which an inertial actuator and collocated velocity sensor are used to control the vibration of a model plate. The Nyquist diagram in Figure 3.10 shows multiple loops on the right hand side, corresponding to plate modes that will be attenuated, and a single loop on the left hand side corresponding to the actuator resonance, whose natural frequency is assumed to be below that of the first plate resonance. This loop will get larger as the feedback gain is increased and ultimately is the cause of instability if the feedback gain is greater than about (Elliott *et al.*, 2001)

$$g_{max} = \frac{2\zeta_a M_m \omega_1^2}{\omega_a},$$ (3.17)

where ζ_a and ω_a are the damping ratio and natural frequency of the actuator, M_m is the moving mass of the structure, which is equal to the modal mass of the plate, $M/4$ in our case, where M is the mass of the plate and ω_1 is its first natural frequency. If the feedback gain is set to half of this maximum value, then the attenuation at the first plate resonance, ω_1, is approximately equal to

$$Attn(dB) = -20\log_{10}\left(\frac{2\zeta_1\omega_a}{\zeta_a\omega_1 + 2\zeta_1\omega_a}\right)$$ (3.18)

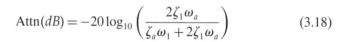

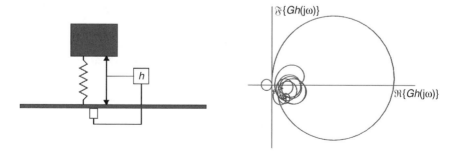

Figure 3.10   An inertial actuator and collocated velocity sensor with local feedback used as a single loop of a decentralised controller on a plate, and the resulting Nyquist diagram showing that while the plate modes result in stable loops with a positive real part, the actuator resonance creates a destabilising loop with a negative real part

where ζ_1 is the damping ratio of the first plate mode before control. In order to get high attenuations with a feedback system using an inertial actuator, it is clear from this equation that the actuator resonance frequency should be well below the first plate resonance, and that the actuator should also be well damped compared with this plate mode. A more quantitative statement of this condition can be obtained by noting that for the control gain with an inertial actuator, Equation (3.17), to be at least as large as the approximate formula for the infinite plate impedance, Equation (3.15), then $4\omega_a/\pi\zeta_a\omega_1$ must be less than unity.

These conditions on the actuator resonance may be difficult to achieve with large engineering structures, for which the first flexible resonance frequency may be at tens of hertz. If the actuator resonance is made too low, the suspension cannot support the weight of the magnet and will bottom out. One method of overcoming this practical limitation of inertial actuators, which has been investigated by Benassi and Elliott (2004), is to incorporate a position sensor, such as a strain gauge on the suspension, within the actuator, and use integral feedback to self-level the actuator so that the inertial mass is 'floating' on a force supplied by a steady current in the coil. It is also possible to incorporate some direct and derivative control within this inner feedback loop to lower the natural frequency and provide additional damping to the actuator, both of which will improve the attenuation in Equation (3.18) above.

An active vibration control system using 24 collocated inertial actuators and velocity sensors on a plate has been investigated theoretically and experimentally by Baumann and Elliott (2006). They show that very good levels of control can be achieved with local, decentralised, controllers, but that the enhancement and stability problems observed for single channel systems due to actuator resonances are even more in evidence for multichannel systems, since the enhancement at the actuator resonance due to one feedback loop affects the open loop response seen by all the other feedback controllers. The importance of good actuator design is thus very clear if decentralised active vibration control systems using large numbers of inertial actuators are to be used in practice.

3.5 LOCAL CONTROL IN THE INNER EAR

The high sensitivity and exquisite frequency selectivity of our hearing has long suggested that there are active amplification processes taking place in the inner ear (Gold, 1948). These active processes, termed the cochlear amplifier, are still not well understood, but it is interesting to compare some of the current models of the cochlea amplifier with the types of active vibration control system discussed above, in order to compare distributed control in a biological system with such control in engineering structures. The snail-shaped cochlea in the inner ear contains two fluid chambers, separated by the basilar membrane, which are mechanically excited at the end by the bones of the middle ear (Pickles, 1988). The basilar membrane is much stiffer near the entrance to the cochlea than it is at the other end, which results in a distribution of natural, or characteristic, frequencies along its length. Figure 3.11 shows a diagrammatic representation of the ear, in which the spiral structure of the inner ear has been straightened out over its length of about 35 mm and the mass and stiffness of the individual parts of the limp basilar membrane are made explicit. The lower part of this figure shows the approximate distribution of characteristic frequencies, along the length of the cochlea in the human ear.

The coupling between the inertia of the fluid and the dynamics of the basilar membrane can, with various simplifying assumptions, be analysed to give a wave equation for the propagation of disturbances within the cochlea, given, for example, by de Boer (1991), which can be written in the frequency domain as

$$\frac{\partial^2 p(x, \omega)}{\partial x^2} - \frac{2j\omega\rho/h}{Z(x, \omega)} p(x, \omega) = -j\omega\rho\, q(x, \omega), \qquad (3.19)$$

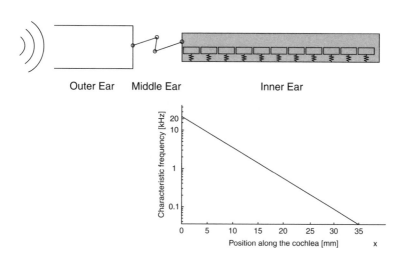

Figure 3.11   Idealised representation of the outer, middle and inner ear, showing the basilar membrane as a series of mass–spring–damper systems distributed down the cochlea, together with the distribution of the natural or characteristic frequencies of these single degree-of-freedom systems

where $p(x, \omega)$ is the complex pressure difference between the two chambers, which are filled with fluid of density ρ, x is the position along the cochlea, h is the height of the chamber and $q(x, \omega)$ is a complex source term. $Z(x, \omega)$ is the mechanical impedance of the basilar membrane alone at position x and frequency ω. The basilar membrane is fairly limp and this impedance is generally taken to be the point impedance of an isolated lumped parameter system. To a first approximation the dynamics at any one point on the basilar membrane alone can be approximated by a single degree-of-freedom, mass–spring–damper system, with a fixed mass, but a stiffness which varies in such a way as to give the distribution of characteristic frequencies shown in Figure 3.11. These individual responses of the basilar membrane segments are then coupled by the inertia of the fluid in the cochlea to give the wave equation in (3.19).

For a given excitation frequency, ω_0, this wave equation has different types of solution for different positions along the basilar membrane compared with the characteristic place, x_0, at which the characteristic frequency, which is generally equal to ω_c, is equal to ω_0. If $x < x_0$, so that $\omega_c > \omega_0$, $Z(x, \omega) \approx k(x)/j\omega$, where $k(x)$ is the stiffness of the basilar membrane at position x, since the basilar membrane is stiffness controlled in this region below resonance. The homogeneous wave equation then becomes

$$\frac{\partial^2 p(x, \omega)}{\partial x^2} + \frac{2\omega^2 \rho}{h\,k(x)}\,p(x, \omega) = 0, \tag{3.20}$$

which has a *propagating* wave solution

$$p(x, \omega) = p_0 e^{-j\omega x/c(x)}, \tag{3.21}$$

where $c(x)$ is the wave speed given by $\left(h\,k(x)/2\rho\right)^{\frac{1}{2}}$. The wave speed falls as x increases, since $k(x)$ decreases along the length of the cochlea, effectively becoming zero at the characteristic place, x_0.

If $x > x_0$, so that $\omega_c < \omega_0$, then $Z(x, \omega) \approx j\omega\,m(x)$, where $m(x)$ is the mass of the basilar membrane at position x, since the basilar membrane is mass controlled in this region above resonance. The homogeneous wave equation then becomes

$$\frac{\partial^2 p(x, \omega)}{\partial x^2} - \frac{2\rho}{h\,m(x)}\,p(x, \omega) = 0, \tag{3.22}$$

which has an *evanescent* solution

$$p(x, \omega) = p_0 e^{-x/\ell(x)}, \tag{3.23}$$

where $\ell(x)$ is the decay length given by $\left(h\,m(x)\big/2\,\rho\right)^{\frac{1}{2}}$, although since $m(x)$ is almost constant, $\ell(x)$ is almost independent of x and has a value of about 1 mm in the human cochlea.

The overall effect is that at a given frequency a wave propagates along the cochlea, with a wavelength that gets smaller and an amplitude that gets larger as the wave speed decreases with increasing x, until the wave stalls at the characteristic place for this frequency and the pressure then rapidly decays away. This gives rise to the characteristic travelling wave envelope widely used to describe the action of the cochlea as a frequency discriminator (e.g. Pickles, 1988). Another way of looking at the dynamic response of the cochlea is to consider the frequency response at one particular position. Figure 3.12 shows the magnitude of the basilar membrane displacement at a position about 17 mm along the cochlea to sinusoidal excitation at various frequencies, simulated using a numerical approximation to the wave equation in Equation (3.19) (Le Henaff *et al.*, 2003). The lower line corresponds to the 'passive' model of the cochlea mechanics outlined above. This gives a heavily damped resonance, which would provide much coarser frequency discrimination than is known to exist at low levels in the healthy cochlea. In order to obtain tuning curves that agree with modern measurements of

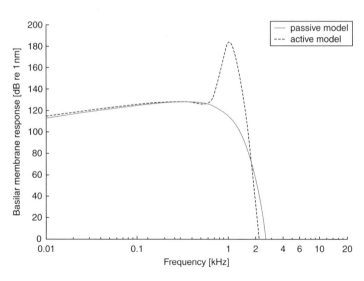

Figure 3.12   Simulated frequency response of the basilar membrane velocity at the mid-point of the cochlea, for the passive, single degree-of-freedom, model of the basilar membrane dynamics (solid) and the active two-degree-of-freedom model (dashed)

the frequency response of physiologically healthy basilar membranes, some kind of active hearing mechanism must be introduced.

There is still considerable controversy about the exact form of this active mechanism, but one model that explains many features found in measurements is that put forward by Neely and Kim (1986). This model accounts not only for the motion of the basilar membrane, but also for the motion of the tectorial membrane, which sits just above the basilar membrane, as shown in Figure 3.13. The shearing between the basilar and tectorial membranes activates the 3500 or so inner hair cells, which are believed to be the main sensory structure in the ear, but is also influenced by the 12000 or so outer hair cells, which are believed to act as both local sensors and fast-acting actuators (Pickles, 1988). The cilia of the hair cells, which protrude into the sub-tectorial gap, are connected to the tectorial membrane in the case of the outer hair cells, while those of the inner hair cells are not connected, and are only driven by the flow of fluid in this gap. Neely and Kim (1986) modelled the overall dynamics of each segment of the active cochlea by the two-degree-of-freedom model also shown in Figure 3.13, in which the lower mass and spring are associated with the basilar membrane and the upper mass and spring model the movement of the tectorial membrane. An additional active force, due to the outer hair cells, acts on the basilar membrane that depends on the relative motion of the basilar and tectorial membranes. The force generated by the outer hair cells, $f(t)$, can be assumed to be approximately proportional to the delayed relative shear displacement, $w(t)$, experienced by their cilia, which connect the tectorial and basilar membranes in Figure 3.13, so that

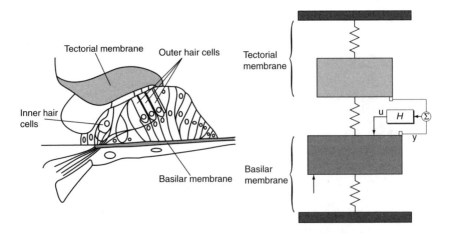

Figure 3.13   The detailed mechanical structure of the cochlear partition (left) and a simplified model of the dynamics, including an active feedback loop (right)

$$f(t) \approx hw(t - \tau), \qquad (3.24)$$

where h is the overall feedback gain and τ is the delay. The dynamic stiffness of the outer hair cell can thus be written in the frequency domain as

$$\frac{F(j\omega)}{W(j\omega)} \approx h\,e^{-j\omega\tau} \approx h - j\omega\tau h, \qquad (3.25)$$

where the second expression assumes that the delay is small compared with a period. The two terms in this final expression can be physically interpreted as a positive stiffness and a negative damper, and it is the latter that supplies power to the system to amplify the motion.

The response of such an active model of the cochlear partition, when incorporated into the wave model above, is also shown in Figure 3.12, and has a much sharper resonance than the passive model.

The additional force in this active model, as generated by the motility of the outer hair cell (OHC) when excited by the shearing motion of the tectorial membrane, forms a feedback loop, as explicitly shown in Figure 3.14. For low-level excitation, the Nyquist plot of the open loop frequency response, for the mechanical constants given by Neely and Kim (1986), gives a series of plots, one of which is illustrated by the solid curve on the right in Figure 3.14. The Nyquist plot passes close to the $(-1, 0)$ point, and this provides the sharp peak in the frequency response seen for this active model in Figure 3.12, which is much more like that measured experimentally at low excitation levels.

For the 40 dB enhancement shown in Figure 3.12, which is typical of the active enhancements that have been measured in the living cochlea (e.g. Johnstone *et al.*, 1986), the polar plot of the open loop frequency response

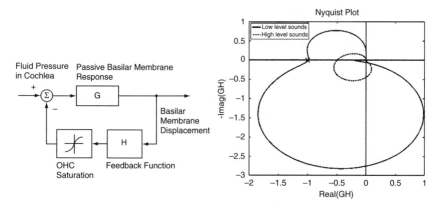

Figure 3.14 Block diagram of the active hearing mechanism as a feedback loop, and the resulting Nyquist plot for low-level excitation

would have to pass so close to the Nyquist point that the multiplicative uncertainty in the open loop response could be no more than 1 %. This implies an extremely delicate setting of the gains for the individual feedback loops formed by each outer hair cell. It is difficult to imagine how the 12 000 outer hair cells, operating independently in this decentralised model, could each have a feedback gain set so close to instability, without there being biological mechanisms to tune the gains and hence regulate the overall amplification.

It is clear that these stabilising mechanisms do not function perfectly, however, since most people have low-level oscillations at various points along their cochlea, which cause measurable tonal pressures in the outer ear under very quiet conditions, known as spontaneous otoacoustic emission (Probst *et al.*, 1991). Such otoacoustic emissions are produced by most primates, in whom the cochlear geometry is slightly inhomogeneous, but are unusual in rodents, in whom the cochlear geometry is extremely regular (Geisler, 1989). A telling exception to this general rule is when there is noise-induced damage to the cochlea, in which case spontaneous otoacoustic emissions are produced at frequencies close to the characteristic frequency of the damaged part of the cochlea (Clark *et al.*, 1984). This all suggests that spontaneous otoacoustic emissions could be due to an abnormal condition of the regulatory mechanisms that exist within a uniform cochlear structure to maintain the individual outer hair cells at maximum amplification and yet inhibit their instability. Little is known about these regulatory mechanisms, which may be biochemical at a very local level, or may involve afferent neural control from the brain stem in a form of outer regulatory feedback loop (Geisler, 1989). Nevertheless, the existence of the limit cycle oscillations that give rise to spontaneous otoacoustic emissions are one of the most direct pieces of evidence for there being active processes in the inner ear.

Some care must be exercised when using linear models such as those described above, however, since even though the dynamics of individual parts of the basilar membrane are stable when considered in isolation, there is no guarantee that the overall model of the different parts of the cochlea, coupled via the fluid chambers, will also be stable. In particular, it may not be obvious using only a frequency-domain analysis for the coupled set of closed loop systems whether the system is stable or not. A complete model of cochlea mechanics is more complicated still, since it is heavily nonlinear. The extremely sharp tuning curves described above are only observed for very low-level excitation of the cochlea (Johnstone *et al.*, 1986). As the sound pressure level increases, the tuning curves become broader, the frequency of the peak response drops and the shape looks much more like the passive response curve in Figure 3.12.

This nonlinearity can be modelled by incorporating a saturation function into the feedback loop shown in Figure 3.14 (Zwicker, 1979; Yates, 1990; Elliott and Harte, 2003), which represents the limited authority of the outer hair cells (as illustrated in Figure 3.21 of Pickles, 1988, for example). At very low sound pressure levels, below about 30 dB SPL, the saturation function acts almost linearly and with a high gain, so that significant positive feedback is present and the response is greatly enhanced. As the sound pressure level rises, the loop gain is gradually reduced by the saturation function, until for very high sound pressure levels, above about 90 dB SPL, the feedback loop has almost no effect and the response of the cochlea is almost passive. The solid Nyquist plot shown on the right hand side of Figure 3.14 shows the situation at low excitation levels, for which the feedback gain is high, and the dashed plot shows the open loop response, at high levels, calculated using the describing function method, for which the effective feedback gain is much lower.

Figure 3.15 schematically shows the level of the BM response as a function of the level of the excitation pressure for the cochlear model with a saturating nonlinearity in the feedback loop, as shown in Figure 3.14. The system is linear at low amplitudes, so that the slope of the level curve is 1 dB/dB and has a gain of about 40 dB. Above an input sound pressure level of about 30 dB SPL, the nonlinearity in the saturation function causes the level curve

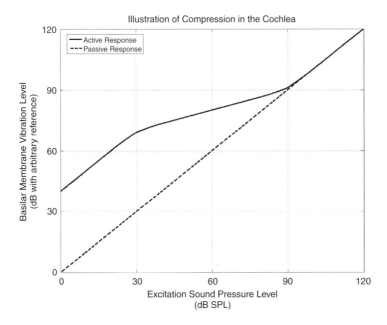

Figure 3.15 Idealised plot of the level of basilar membrane (BM) velocity against exciting sound pressure level for a feedback loop with a saturating nonlinearity

to be compressed, so that its slope is about 1/3 dB/dB in this case. For input sound pressure levels above about 90 dB SPL, however, the feedback gain is so low that the response is almost passive, in which case the cochlea responds linearly with a level curve slope of 1 dB/dB. The main features of this level curve are in good agreement with those measured experimentally (Johnstone *et al.*, 1986). Between input sound pressure levels of about 30 dB SPL and 90 dB SPL, a range of 60 dB, the BM response only increases by about 20 dB, which is one of several mechanical compression mechanisms that are present in the ear, the acoustic reflex of the middle ear being another (Pickles, 1988). Overall, these mechanisms compress the dynamic range of pressures over which we can hear, about 120 dB, into the more limited dynamic range over which the inner hair cells can respond with reasonable fidelity, which is about 40 dB. Another manifestation of the mechanical nonlinearity in the cochlear amplifier is the generation of intermodulation distortion. This is perceived as a 'combination tone' when two sinusoids are heard at one time and gives rise to distortion product otoacoustic emissions (Pickles, 1988).

Although we do not have a complete understanding of the active processes within the ear, it seems clear that the cochlear amplifier is an example of a distributed biological control system, whose gain is somehow regulated to be just at the point of instability under quiet conditions, when the maximum sensitivity is required.

3.6 CONCLUSIONS

Centralised systems can perform well in selectively controlling vibration, but their stability is potentially sensitive to transducer failures, they also require a good deal of rapid communication and their complexity increases rapidly with the size of the system. In a fully decentralised control system only local feedback loops are implemented, and if the actuator and sensor are both dual and collocated, constant gain feedback controllers are guaranteed to be stable, even if the response of the structure changes or if some of the transducers fail. The actuator, sensor and controller for each loop could also be implemented as a single module, which could be mass produced, and only the number of these identical modules would need to be increased to control larger systems. Although decentralised active vibration controllers using collocated force actuators and velocity sensors are not selective, they are able to control the global behaviour of the system almost as well as centralised systems using the same actuators and sensors

Practical decentralised vibration control systems are generally not as straightforward as this, since it is not possible to implement ideal force actuators without another structure to react off Piezoceramic actuators have been used in practical feedback loops, but require careful control of their high-frequency response if stability is to be preserved. Inertial electromagnetic actuators offer another practical solution in some applications, although to give the best performance their resonance must have a low natural frequency and be well damped, particularly in systems with many channels.

As a demonstration of the decentralised control approach in nature, the active process within the inner ear is considered as a distributed control system. Although we do not have a good knowledge of the exact mechanisms of the active hearing process that makes our ears so acute, it is clear that the outer hair cells play an important role. These cells can respond mechanically to an electrochemical excitation with a bandwidth of tens of kilohertz, and are thought to generate nonlinear feedback loops acting on the basilar membrane, tuned close to instability to amplify its response. The high bandwidth is possible because of the small scale of the active structure, of the order of a few micrometres, and because the loops only act locally, avoiding propagation delays. At very low levels of excitation these loops appear to be regulated to be very close to the point of instability, so that some of them are not stable and have limit cycles that give rise to spontaneous otoacoustic emissions in most people's ears. Very little is known about this regulatory mechanism and it may be that by studying this process more will be learnt about how to tune distributed active control systems for engineering structures. It is, nevertheless, clear that the inner ear is a remarkable natural example of an adaptive structure.

ACKNOWLEDGEMENTS

I have had the benefit of many stimulating conversations during the preparation of this chapter, and I would particularly like to acknowledge Paolo Gardonio and Olie Baumann for sharing with me their results on active vibration control on plates and Robert Pierzycki and Emery Ku for some of the results on cochlear modelling.

REFERENCES

M.J. Balas (1979) Direct velocity feedback of large space structures. Journal of Guidance and Control, **2**, 252–253.

O.N. Baumann and S.J. Elliott (2006) Decentralised control using multiple velocity feedback loops with inertial actuators. *Proc. ACTIVE 2006*, Adelaide, Australia.

L. Benassi and S.J. Elliott (2004) Active vibration isolation using an inertial actuator with local displacement feedback. Journal of Sound and Vibration, **278**, 705–724.

R.J. Benhabib, R.P. Iwens and R.L. Jackson (1981) Stability of large space structure control systems using positivity concepts. Journal of Guidance and Control, **4**, 487–493.

A.J. Bullmore, P.A. Nelson, A.R.D. Curtis and S.J. Elliott (1987) The active minimisation of harmonic enclosed sound fields, Part III: A computer simulation. Journal of Sound and Vibration, **117**, 15–33.

R.L. Clark, W.R. Saunders and G.P. Gibbs (1998) *Adaptive Structures, Dynamics and Control*, John Wiley & Sons, Ltd.

W.W. Clark, D.O. Kim, P.M. Zarek and B.A. Bohne (1984) Spontaneous otoacoustic emission in chinchilla ear canals: Correlation with histopathology and suppression by external tones. Hearing Research, **16**, 299–314.

L. Cremer and M. Heckl (1987) *Structure-borne Sound*, Springer-Verlag (translated and revised by E.E. Ungar).

E. de Boer (1991) Auditory physics. Physical principles in hearing theory III. Physics Report, Review Section of Physics Letters, **203**(3), 125–231.

S.J. Elliott and J.M. Harte (2003) Models for compressive nonlinearities in the cochlea. ISVR Technical Memorandum 913, University of Southampton.

S.J. Elliott, M. Serrand and P. Gardonio (2001) Feedback stability limits for active isolation systems with reactive and inertial actuators. ASME Journal of Vibration and Acoustics, **123**, 250–261.

S.J. Elliott, P. Gardonio, T.C. Sors and M.J. Brennan (2002) Active vibroacoustic control with multiple local feedback loops. Journal of the Acoustical Society America, **111**, 908–915.

W. Engles, O.N. Baumann, S.J. Elliott and R. Fraanje (2006) Centralised and decentralised control of structural vibration and sound radiation. Journal of the Acoustical Society of America, **119**, 1487–1495.

W.P. Engles (2006) Decentralised velocity feedback control of structures. PhD Thesis, ISVR, University of Southampton.

M.I. Friswell, S.D. Garvey, J.E.T. Penny and A. Chan (1996) Selection of control topology for decentralised multi-sensor multi-actuator active control. *Proc. 14th Modal Analysis Conference*, Dearborn, USA.

C.R. Fuller, S.J. Elliott and P.A. Nelson (1996) *Active Control of Vibration*, Academic Press.

P. Gardonio, E. Bianchi and S.J. Elliott (2004) Smart plate with multiple decentralised units for the control of sound transmission. Part I: Theoretical predictions, Part II: Design of the decentralised control units, Part III: Control system implementation. Journal of Sound and Vibration, **274**, 163–232.

W.K. Gawronski (1998) *Dynamics and Control of Structures: A Modal Approach*, Springer-Verlag.

D.C. Geisler (1989) *From Sound to Synapse: Physiology of the Mammallian Ear*, Oxford University Press.

T. Gold (1948) Hearing II, the physical basis of the action of the cochlea. Proceedings of the Royal Society, **135**, 492–498.

N. Hirami (1997) Optimal energy absorption as an active noise and vibration control energy strategy. Journal of Sound and Vibration, **200**, 243–259.

X. Huang, S.J. Elliott and M.J. Brennan (2003) Active isolation of a flexible structure from base vibration. Journal of Sound and Vibration, **263**, 357–376.

B.M. Johnstone, R. Patazzi and G.K. Yates (1986) Basilar membrane measurements and the travelling wave. Hearing Research, **22**, 147–153.

D. Karnopp (1995) Active and semi-active vibration isolation. ASME Journal of Mechanical Design, **117**, 177–185.

S.-M. Kim, S.J. Elliott and M.J. Brennan (2001) Decentralized control for multi-channel active vibration isolation. IEEE Transactions on Control Systems Technology, **9**(1), 93–100.

B. Le Henaff, S.J. Elliott and C. Maury (2003) Modelling wave propagation in the cochlea. ISVR Technical Memorandum 925, University of Southampton.

L. Meirovitch (1990) *Dynamics and Control of Structures*, John Wiley & Sons, Inc.

S.T. Neely and D.O. Kim (1986) A model for active element in cochlear biomechanics. Journal of the Acoustical Society of America, **79**(5), 1472–1480.

P.A. Nelson (1996) Acoustical prediction. *Proc. InterNoise 96*, Liverpool, UK.

B. Petitjean and I. Legrain (1996) Feedback controllers for active vibration suppression. Journal of Structural Control, **3**, 111–127.

J.O. Pickles (1988) *An introduction to the physiology of hearing*, 2nd edition, Academic Press.

A. Preumont (2002) *Vibration Control of Active Structures, an introduction*, 2nd edition, Kluwer Academic.

R. Probst, B.L. Lonsburg-Martin and G.K. Martin (1991) A review of otoacoustic emissions. Journal of the Acoustical Society of America, **89**, 2027–2067.

W. Redman-White, P.A. Nelson and A.R.D. Curtis (1987) Experiments on the active control of flexural wave power. Journal of Sound and Vibration, **112**, 187–191.

M. Serrand and S.J. Elliott (2000) Multichannel feedback control for the isolation of base-excited vibration. Journal of Sound and Vibration, **234**(4), 681–704.

S. Skogestad and I. Postlethwaite (1996) *Multivariable Feedback Control*, John Wiley & Sons, Inc.

J.Q. Sun (1996) Some observations on physical duality and collocation of structural control sensors and actuators. Journal of Sound and Vibration, **194**, 765–770.

W. Variyart, M.J. Brennan and S.J. Elliott (2002) Active damping for a beam using feedback control. *Proc. ACTIVE 2002*, Southampton, UK, pp. 839-850.

G.K. Yates (1990) Basilar membrane nonlinearity and its influence on auditory nerve rate-intensity functions. Hearing Research, **50**, 145–162.

E. Zwicker (1979) A model describing nonlinearities in hearing by active processes with saturation at 40 dB. Biological Cybernetics, **35**, 243–250.

4
Lightweight Shape-Adaptable Airfoils: A New Challenge for an Old Dream

L.F. Campanile

Swiss Federal Laboratories for Materials Testing and Research (EMPA), Dübendorf, Switzerland

4.1 INTRODUCTION

Recently the second centenary of the birth of Isambard Kingdom Brunel, one of the most talented engineers of the nineteenth century, was celebrated. His visions shaped the transportation system of the modern world in a lasting way, and his profound knowledge of structural engineering provided a substantial contribution to the development of the science of lightweight mechanical systems.

Brunel's lifework is probably one of the best documented of its kind, owing to his meticulous notes as well as to his letters, to the loving attention of many members of his family in collecting and archiving documents and making them available to the public and, finally, to the efforts of excellent historians like L.T.C. Rolt (1989) or R. Angus Buchanan (2002), who worked

Adaptive Structures: Engineering Applications   Edited by D. Wagg, I. Bond, P. Weaver and M. Friswell
© 2007 John Wiley & Sons, Ltd

through the huge amount of material and undertook the task of composing Brunel's biographies.

The detailed documentation about Brunel's professional life not only high-lights his great achievements, but also makes a fair account of some spec-tacular failures which marked his career. These 'disasters' – to use the term chosen by Buchanan – do not in any way belittle the value of Brunel as an engineer, but rather put him into a more human dimension, by stressing that risk is somehow a corollary to courage and creativity, and surely contribute to rendering Brunel an example for future generations of engineers. Besides this human dimension, also Brunel's connotation as a visionary allows for a positive interpretation of these unlucky events: being a visionary implies, after all, some sort of predisposition to disasters, in particular to the special sorts of disasters which are not imputable – or, at least, not completely imputable – to wrong ideas but rather to right ideas which are ahead of their time.

An interesting example in this context is the *atmospheric railway* adven-ture. The idea behind the atmospheric railway consisted in propelling the train – without a locomotive – by connecting the front carriage to a piston which was free to slide inside a tube located between the rails. Evacuation of the tube in front of the piston produced the thrust force which finally accelerated the train. Not only Brunel but also the passengers who used it for about a year starting from September 1847 were enthusiasts of trains without smoke, dirt or loud noise, and with the impressive acceleration resulting from the absence of a heavy locomotive (Hart-Davis, 2006). Unfortunately, the valve permitting free motion of the connection arm between carriage and piston along the line – while keeping the longitudinal slot sealed when the arm was not passing through – did not operate correctly, and no feasible alternative solution was found. As a consequence, the system was abandoned and ended up as a financial catastrophe. Anyway, as Buchanan points out, the atmospheric railway system represents somehow a precursor of modern electric traction, which provides continuous power supply along the track as well. In this sense, Brunel's vision incorporated a correct intuition which just did not match the technological state of progress of the times in which it was generated.

At the time when Brunel began to conceive his vision of a modern ground and sea transportation system, another vision had fascinated scientists and engineers for at least three centuries:[1] the vision of flight. By the dawn of the twentieth century, nearly 100 years after the appearance of the first

---

[1] It is not easy to determine the point in history at which the dream of flight leaves a mere mythological and legendary dimension and enters the world of engineering. Here we elect to consider Leonardo da Vinci (1452-1519) as the first aeronautical engineer.

locomotive and 65 years after Brunel's *Great Western* completed its maiden
trip from Bristol to New York, powered flight finally managed to become
a reality, and within a further 50 years at most the aircraft took its firm
place in the framework of long-distance travel beside the train and ship. The
original vision, however, i.e. that of bio-mimetic flight, did not come true: the
dream of a flying machine could only be realised when engineers abandoned
the idea of flying in the same way a bird does. Still, the fascination of
the silent, efficient, reliable 'natural way' of flying survives, and the desire
of reproducing it technically exists in our times. Thus, while bio-mimetic
flight is commonly considered as the wrong approach to aeronautics, the
question arises whether the idea was erroneous or, once again, just at the
wrong time.

Some noticeable trends of research in aerospace, or more generally in
applied mechanics, give strength to the latter alternative. We will examine
them in the course of this chapter and discuss the underlying philosophy,
whose implications extend beyond the special issue of bio-inspired flight
concepts and are likely to have a large impact on the design and optimisation
of mechanical systems in general.

4.2   OTTO LILIENTHAL AND THE FLYING MACHINE AS A SHAPE-ADAPTABLE STRUCTURAL SYSTEM

While comparing the ways a bird and a fixed-wing aircraft fly, two
essential differences appear. The first, and more evident, concerns the
way of producing thrust: while birds achieve it by a combined flapping–
rotation motion of the wings, aircraft are equipped with separate propulsion devices. The second difference, perhaps more subtle, involves how
the resultants of the aerodynamic forces are regulated in order to control
the flight path and attitude: smooth, virtually unrestricted modifications of
the airfoil geometry versus the activation of a few rigid control surfaces.
As will be shown in the following, these two differences are strongly
related to another and can even be regarded as two aspects of the same
thing: bird flight – or, more generally, animal locomotion – is based on
highly coupled shape-adaptable systems while aviation – or, correspond-
ingly, human-made transportation devices – essentially rely on modular
systems whose elements are in charge of clearly separated (or, at least,
weak coupled) mechanical tasks. The distinction itself between lift, drag,
thrust and control forces arises from this task separation principle and is
not necessary while describing flight in nature. In his book on bird flight
(1889), Otto Lilienthal approaches flight as the problem of 'processing the

surrounding air with properly shaped wings in a proper way'. Impressive in its simplicity, this definition suggests a quite revealing abstraction: the task of realising a flying machine can be essentially reduced to the task of realising a proper shape-adaptable structural system. Not only birds or other flying animals, but also most aircraft interact with the flow 'just' through proper geometry changes (this includes thrust produced by propellers and rotors and can be partially applied to turbofan engines too). According to this abstraction, differences between artificial and natural flyers can be related to corresponding differences in geometry management and the challenge of emulating biological flight essentially reduces to the complexity of reproducing the shape adaptation mechanisms which are present in nature.

In the framework of the synthesis of a shape-adaptable structural system, mechanical design deals with the task of choosing the properties and distribution of load-carrying material and actuators in order to fulfil:

- a set of *deformability* requirements specifying the geometrical changes which the system under consideration must be able to perform;
- a set of *stiffness* requirements which define the allowed deviations from the desired shape under given loads;
- *strength* requirements which specify the loads to be carried by the system without damage;
- *activability* requirements which state that the desired deformation is achievable by loading through the actuator system; and possibly
- a set of further requirements – mostly coupled with another and related to possible payload requirements – which define boundaries for the system's *weight* and *energy consumption*.

Of course, mechanical design as defined above does not constitute a closed task since the exact definition of the single requirements is not a priori available, but is somehow part of the unknowns of the whole synthesis problem. For this reason, mechanical issues cover only a limited portion of the complexity involved in the global design process (essential questions like, for instance, sensor and control design or energy reserve allocation are not considered in this context). Still, the issue of mechanical design is highly representative, since it concerns components with a strong impact on the overall system weight and therefore involves some crucial challenges of bio-mimetic flight. On this basis, and keeping the focus on problems which are specifically mechanical, the key differences between the technical and the biological approach to flight can be more easily understood and it can be made clear why the former is going to get increasingly closer to the latter in the course of progress.

Before moving to a more detailed discussion on task separation as the key principle of conventional mechanical design, a last remark on the different sorts of bio-mimetic measures is useful. As mentioned before, the distinction between thrust production and flight control measures which was introduced at the beginning of this section is not necessary and somehow arbitrary. Above all, a bird in steady flight just manages to obtain, from its interaction with the flow, a resulting force which equates to gravity; the horizontal component of this force is therefore zero and it is not clear how to split it into a thrust and a drag force which equilibrate one another. Still, if necessary, a distinction of an energetic kind between thrust production and flight control measures can be made, since the former continuously transfer mechanical power from an internal energy reserve to the flow by means of a cyclic motion, while the latter are typically non-cyclic and do not involve a substantial power flow. Such a classification can be useful, particularly while figuring out the implementation times of bio-mimetic concepts in aircraft: the replacement of control surfaces through extended shape-control capabilities is likely to require less time than the replacement of conventional engines through bio-mimetic thrust production devices. Finally, our classification can be completed by a third group of bio-mimetic features, which was not considered above since it has no counterpart in state-of-the-art aircraft: the capability of quasi-statically adapting the airfoil geometry with the aim of reducing off-design effects. As far as the time scale is concerned, this third group of measures can be reasonably placed on the short-term range, prior to flight control.

4.3   SIR GEORGE CAYLEY AND THE TASK SEPARATION PRINCIPLE

Just a few years before Isambard Kingdom Brunel was born, another illustrious British engineer, Sir George Cayley, had set a milestone on the way towards the realisation of powered flight. In asserting the importance of separating the tasks of lift and thrust production and assigning them to different subsystems of an aircraft, he formulated a principle which is today still considered as one of the keys to airplane design (Campanile, 2006). The actual merit of Cayley was to propose a pragmatic solution to the problem of flight and shorten in this manner the way toward a feasible concept. In doing this, however, he narrowed the field of the available solutions: for instance, vertical take-off and landing aircraft cannot be realised on the basis of separated lift and thrust production devices (Staufenbiel 2002).

On closer inspection, Cayley's idea of lift and thrust separation appears as the expression of a more general (and surely older) principle of mechanical design: the synthesis of a complex mechanical system (typically with multiple interactions among different physical effects and conflicting requirements) can often be successfully approached by breaking up the system to be designed into separate (or weakly coupled) units; however, since this corresponds to a radical – and somehow arbitrary – restriction of the design space, such an approach usually suffers from poor optimisation potential.

As far as the mechanical design of shape-adaptable systems is concerned, the implications of this task separation principle are far reaching and not limited to just the separation of lift and thrust production devices. Let us consider one of the classic shape adaptation problems: the realisation of airfoils with changeable camber for the purpose of regulating lift. A substantial conflict occurs between the deformability requirements, defining the change in camber which must be allowed by the airfoil, and the stiffness requirement, which limits shape changes under load. The conventional solution based on an actuated flap (see Figure 4.1) implements an almost complete separation between the task of producing the desired geometry changes and that of carrying loads. A rigid-body mechanism (the flap and the fixed part of the airfoil, mutually connected by a hinge) guarantees unrestricted deformability in one single degree of freedom; all the remaining degrees of freedom can then be provided with sufficient load-carrying capabilities (strength and stiffness) without interacting with the system's deformability. In this way, the conflict between load-carrying capabilities and deformability is restricted to the single degree of freedom of the rigid-body mechanism and can be easily solved by providing the actuator with proper strength and (active or passive) stiffness. By coupling a larger number of degrees of freedom and harmonising the contribution of active and passive elements, alternative solution are possible, as nature suggests. Such systems show typically a better performance (in particular, smoother geometry changes and weight savings through load distribution) but are definitely more complex to design.

The key element of the conventional solution for changeable camber, the revolute joint, is also the basis of an invention which somehow represents the

Figure 4.1 A conventional airfoil with changeable camber

epitome of technical progress (Gould, 1981): the wheel. In a certain sense, the wheel implements the principle of separation between lift and thrust in a device for terrestrial locomotion: it provides a concept to support the payload – by equilibrating gravity – without restraining horizontal motion. In the plane of rotation, it offers virtually infinite stiffness in one direction and zero stiffness in the direction normal to it. In doing this, it allows for a decoupled design in which the propulsion unit and the carriage of a terrestrial vehicle can be designed separately. But, even in the prominent case of the wheel, these practical advantages have their counterpart in a poor optimisation potential: wheel-based transportation concepts are not superior, in a general sense, to other solutions, which reasonably explains why they are not present in nature. The interested reader can refer to Gould (1981) and La Barbera (1983) for a thorough discussion on this topic and find some further suggestions in Campanile (2006). Still, a few additional considerations on this point should be useful here. Let us consider the problem of designing a shape-adaptable system for terrestrial locomotion, i.e. a device which by means of active geometry changes on its surface and the resulting interaction with the ground should be able to perform a controlled motion on the earth surface. By choosing the option of a vehicle with powered wheels – as compared, for instance, with a bio-inspired concept based on leg motion (quadruped) or on travelling waves (worm) – significant disadvantages are encountered on unprepared terrain: a powered wheel, for instance, is not able to surmount an obstacle higher than the wheel radius (Bekker, 1956). On the other hand, the invention of the wheel stems – in all probability – from the mere passive task of producing an energetically efficient solution for animal traction. Since, in this case, task separation is not just an option, but a requirement of the problem, the choice of a decoupled solution appears more obvious. This is not the case in aeronautics: while approaching the design problem as a whole, engineers and scientists were particularly receptive to bio-mimetic suggestions, and the separation of mechanical tasks needed to be deliberately countered to the coupled approach of nature as a pragmatic choice. What appears to be a corollary of the wheel principle would have perhaps been recognised and applied long before Cayley if animal-powered flight had ever been a realistic option.

4.4   BEING LIGHTWEIGHT: A CRUCIAL REQUIREMENT

Shape control by means of rigid-body mechanisms was not the standard solution for load control in the first years of aeronautics. Early concepts, like the one implemented in the famous wing-warping device of Orville

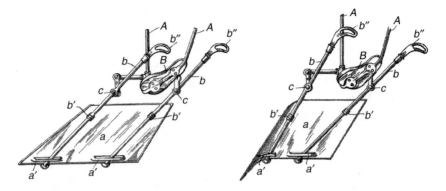

Figure 4.2   An early concept for a control surface (Schmidt, 1895)

and Wilbur Wright's *Flyer* (Wright, 1988a), were based on the exploitation of structural flexibility instead. While in most cases – see for instance the device represented in Figure 4.2 – bio-mimetic suggestions, or simply the aerodynamic advantages expected by a smooth deformation pattern, should have played a major role, in the case of the Wrights' wing-warping device (see Figure 4.3) the choice of a concept based on elastic deformation was essentially dictated by structural considerations. As Orville Wright reports in a paper written in connection with a lawsuit, the original concept for a roll control device was, indeed, based on rigid-body mechanisms. Later on, however, it was abandoned because the Wright Brothers 'did not see any method of building this device sufficiently strong and at the same time light enough' (Wright, 1988b). Using elastic strain instead of relying on moveable parts was hence a deliberate response to a conflicting requirements scenario.

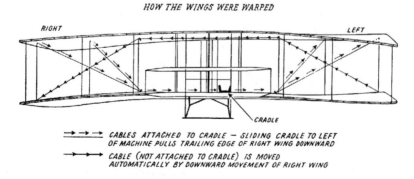

Figure 4.3   The wing-warping device of the Wrights' *Flyer* (Wright, 1988a). Reproduced by permission of Dover Publications, Inc.

Orville Wright's assertion makes the point – and it probably represents the first account in the literature in this sense – on a central fact: shape-adaptable systems based on rigid-body mechanisms suffer from considerable weight penalties.

Conciliating the conflicting requirements of deformability, load-carrying capability and low weight is the dominant issue in the mechanical design of shape-adaptable systems. While established technologies and disciplines which successfully deal with the fulfilment of just two out of the three requirements are presently available, the simultaneous accomplishment of all three of them represents a real challenge. Filling this gap is likely to keep busy a large number of researchers in the near future.

Even though the coupling with the actuator system can play an important role in this context, in particular with regard to stiffness aspects, the above-addressed challenge concerns primarily the passive part of the system, and therefore we will focus on it in the remaining part of this section. Let us now look at the diagram in Figure 4.4. The first 'binary' discipline is lightweight design, which covers a large spectrum of procedures, criteria and techniques for the analysis and synthesis of structures with a high strength-to-weight ratio. Lightweight structures, however, are typically provided with a low degree of deformability. This is a direct consequence of the key principle of lightweight design, which tends to realise a stress state with mainly membrane loading and to avoid bending and torsion loading of single components. Under certain conditions, maximising stiffness is equivalent to minimising weight and this criterion is often used as the basis of structural optimisation procedures. The second discipline which is capable of joining two of the three above-mentioned requirements is machine engineering, or, in particular, the science of conventional mechanisms. They are

Figure 4.4   The requirement triangle of lightweight shape adaptation

highly deformable and able to carry high loads, but, as mentioned earlier and as discussed below in more detail, far from being lightweight optimised. Finally, a third specialty deals with the remaining couple of requirements: the science of compliant systems, i.e. systems designed to reach a given level of deformability by exploiting structural strain. They offer an excellent combination of deformability and low weight, but are limited – at the present state of the art – in their load-carrying capability: systems with lumped compliance, provided with cut-outs or other kinds of solid-state hinges, typically suffer from stress concentration effects in the compliant regions, whereas systems with spread compliance are more difficult to control owing to their low stiffness in a large number of degrees of freedom. Compliant systems will also be addressed later in more detail, since they are likely to provide the basis for the future development of light, high-load, high-deformability systems.

Returning to the historical evolution of control surfaces, it is clear from the above classification that early concepts, like the wing-warping device of the *Flyer*, must have been susceptible to load problems. Indeed, as soon as aviation ceased to be a leisure-time activity practised in favourable weather, at low speed and with minimum payload (Campanile, 2005) and entered the more serious business of World War I, aircraft designers converted to pin-jointed control surfaces. Weight penalty problems had evidently become tolerable – or at least less important than load problems – but had to be maintained at a low level by keeping the number of articulations to a minimum. This, in turn, limited the number of controllable degrees of freedom and rendered smooth deformation patterns no longer practicable.

In the course of the history of aeronautics and all the way till today, a large number of inventions were elaborated which aimed at realising smooth shape changes as well as at increasing the number of controllable geometry degrees of freedom on the basis of conventional mechanisms, but they generally turned out not to be feasible. Still, the resulting patent literature offers an excellent insight into how significant this issue has always been to aircraft engineers and which kinds of advantages were expected from airfoils with enhanced geometry management.

One of the first inventions in this direction was patented by the British *Varioplane Company* (Holle and Judge, 1916) with the aim of regulating lift in a more efficient way by realising a smooth camber variation in the airfoil as opposed to a mere change in the angle of attack. The invention essentially consists in a trailing-edge flap whose activation is coupled with a device situated in the fore part of the wing. By the action of this mechanism and by an open, flexible shell which surrounds the whole apparatus, a downward deflection of the flap increases the camber of the upper surface of the wing in

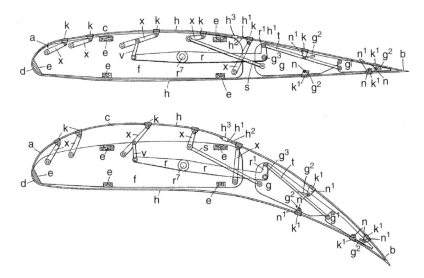

Figure 4.5 Variable-camber concept by Holle and Judge (1916). The UK Patent Office © Crown copyright 1916

its leading region (see Figure 4.5). The opposite sign of the work performed by the aerodynamic loads on the leading and on the trailing region of the airfoil leads to a sort of energy balance which reduces the control effort.

A similar concept was elaborated later by Rocheville (1932). It also makes use of a conventional, rigid, trailing-edge flap and of a trussed mechanism in the fixed part of the wing (see Figure 4.6). In 1924, Antoni developed a solution for a variable-camber airfoil (Antoni, 1928) with the purpose of adapting the airfoil camber to different flight conditions (reducing off-design effects, see end of Section 4.3). It is essentially based on articulated ribs

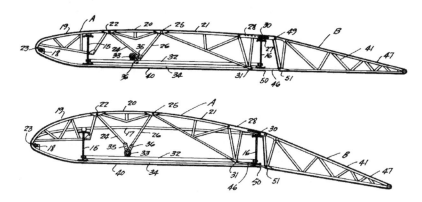

Figure 4.6 Variable-camber concept by Rocheville (1932)

consisting of two rigid segments each, which shape a flexible shell and produce in this way a smooth deformation pattern.

The airfoil design proposed by Parmele in 1928 allows changes in the section thickness 'to impart a greater lifting capacity to the heavier than air flying machine during the initial or starting flight and to provide means for the air ship attaining greater speeds after the machine has risen sufficiently from earth' (Parmele, 1931). The shape changes are realised by means of a segmented rib whose elements, interconnected by means of revolute hinges or linear slide hinges, directly define the upper contour of the wing (see Figure 4.7). A further concept of the early decades of aviation, devised by Grant in 1933, is shown in Figure 4.8. It employs a system of triangular truss panels to adapt the airfoil camber (Grant, 1935).

In the second half of the twentieth century the search for a feasible solution continued with undiminished intensity. Lyon's 'variable shaped airfoil' (1965) realises smooth section changes by means of flexible ribs floatingly connected to the spars by pivoted members (see Figure 4.9). The purpose of the invention is manifold and includes load alleviation ('an airfoil structure

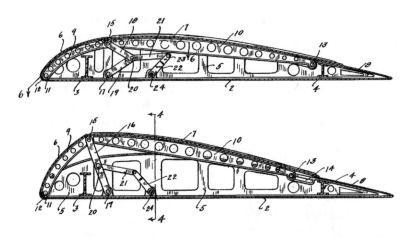

Figure 4.7   Variable-camber concept by Parmele (1931)

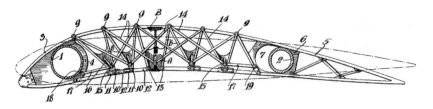

Figure 4.8   Variable-camber concept by Grant (1935)

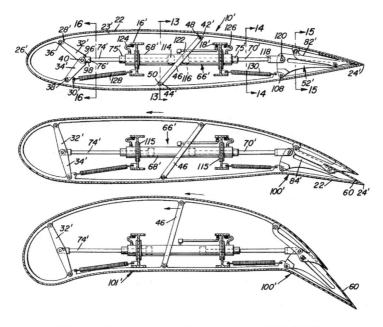

Figure 4.9   Lyon's 'variable shaped airfoil' (1965)

which may give or flex when subjected to heavy stresses or shocks resulting from abrupt changes in ambient atmospheric conditions'), manoeuvrability enhancement, drag reduction, de-icing and even pitch trim by means of an adjustable centre of pressure.

In several solutions which were presented more recently, segmented mechanisms are employed for the purpose of adjusting the airfoil camber. Pierce's design (1978) relies on a supporting chain whose members are mutually jointed, with the joints located in the neighbourhood of the lower surface of the airfoil. The upper surface is defined by a flexible skin which is free to slide on the chain members. By mutually rotating the members of the supporting chain the lower side of the airfoil section is deformed to a polygonal contour, while the upper side is smoothly bent (see Figure 4.10). Berry's device (2000) produces camber changes in a similar way but uses a more complex kinematic chain with just one degree of freedom: the relative motion between two main links forces the remaining member of the chain, which finally describes the airfoil shape, into a defined motion. An analogous principle is used by Piening and Monner (2000) but with a different layout of the slide and revolute bearings, which allows the upper and the lower side of the airfoil section to be similarly deformed.

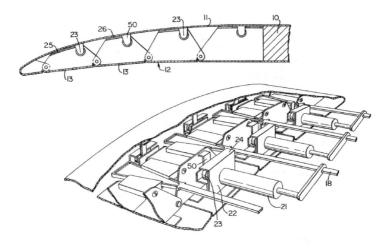

Figure 4.10   Segmented design for a variable-camber airfoil by Pierce (1978). The UK Patent Office © Crown copyright 1978

Rigid-body mechanisms of more complex kind were adopted by Lee and Hanlon (1977) and McKinney (1982, see Figure 4.11) to realise large camber changes in airfoils. Further inventions which are worth mentioning are the variable-camber devices of Ash and Walley (1979), Eppler (1985), McKinney (1977), Statkus (1982) and Rowarth (1981), all of them implementing a supporting rigid-body mechanism and a flexible skin or sliding cover to build the airfoil surface.

Even if weight penalty effects in conventional mechanisms are of fundamental importance and evidently played a key role in driving the state of the

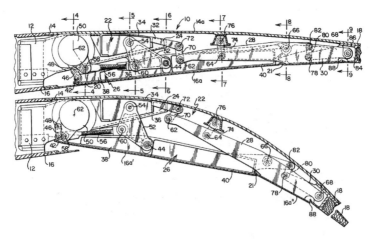

Figure 4.11   Variable-camber trailing edge by McKinney (1982)

art of airfoil shape management to the point at which it presently is, they are only marginally addressed in the literature. A basis for the quantitative analysis of such effects was proposed by the author (Campanile, 2004) and applied to the case of pin-jointed mechanism with or without rolling bearings. The key findings of this study are summarised in the following.

The central issue of this analysis is to evaluate the price which must be paid – in terms of additional weight – to provide a structural element (e.g. a prismatic bar) with a revolute hinge, under the condition that the hinged element is able to carry the same load as the fixed-geometry component. Of course, the comparison of the two constructions is to be made on a common basis: the geometrical configuration of the articulated element must reproduce the geometry of the original structure and the load must be chosen in such a way that mechanism degrees of freedom are not accelerated. In the case of the prismatic bar and of the corresponding articulated mechanism (see Figure 4.12) in the plane of rotation only pure tension/compression loading or pure shear loading is allowed for the comparison, as well as any linear combination of these two load cases.

Under these conditions, a measure for the occurring weight penalty is given by dividing the volume of the additional material required by the hinge by a reference volume:

$$\alpha = \frac{\Delta V}{V_0}, \tag{4.1}$$

the reference volume V_0 being, in turn, related to the load-carrying capability of the fixed-geometry structure. If A denotes the cross-section of the prismatic bar represented in Figure 4.12 (dimensioned for the load under consideration), the reference volume is

$$V_0 = A^{3/2}. \tag{4.2}$$

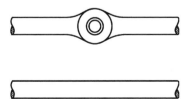

Figure 4.12   A pin-jointed articulation and the corresponding reference element with fixed geometry

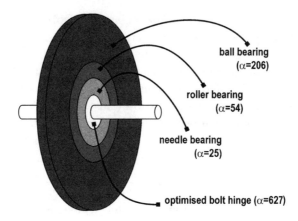

Figure 4.13   Weight penalty for conventional hinges

Values for this weight penalty factor range from 6.27 for a bolted joint with tight clearance (Campanile, 2005) to over 200 for ball-bearing hinges (Campanile, 2004). Figure 4.13 illustrates the physical meaning of these results: introducing, for instance, a ball-bearing joint in a cylindrical rod subject to an axial load leads to a weight penalty effect which is equivalent to providing the rod with a disc plate (made of the same material) whose thickness is nearly the same of the rod (exactly $\sqrt{\pi}/2 = 0.89$ times the rod thickness) and whose diameter is equal to $\sqrt{\alpha} = 14.4$ times the rod diameter.

4.5 COUPLING MECHANISM AND STRUCTURE: COMPLIANT SYSTEMS AS THE BASIS OF LIGHTWEIGHT SHAPE-ADAPTABLE SYSTEMS

4.5.1 The Science of Compliant Systems

After the large number of unsuccessful attempts to adapt the technology of conventional mechanisms to the targets of lightweight shape control it is natural to focus on the other approach to deformability: the use of compliant systems.

As mentioned earlier, compliant systems (also usually called compliant *mechanisms*) rely on elastic strain in order to reach the desired degree of deformability. Of course, every elastic structure is subject to deformations under load and is therefore provided with some compliance; but only a special category of systems is designed to this purpose. Hence, a compliant system can be defined as a continuum – made of an elastic material – which explicitly fulfils a deformability restriction. It is somehow reciprocal to a structure, defined as a material system able to fulfil explicitly a strength restriction (i.e. able to carry a defined load without damage), since deformability is the counterpart of strength in a prescribed-displacement view (as opposed to the prescribed-force view usually applied to structures). To avoid a common source of confusion, it is important to point out here that deformability is not the same as flexibility: the former defines the limit for the deformation which the mechanical system can undergo without being damaged, but says nothing about the forces needed to reach such deformation; the latter – as the reciprocal of stiffness – provides information about the displacements of the system under a given load. To give an example, a beam with a rectangular cross-section can be provided with a given deformability in bending by limiting its thickness, whereas stiffness can be indefinitely increased acting on the beam width. Incidentally, the fact that deformability and stiffness do not exclude each other plays an important role in the design of shape-adaptable systems (which have to provide both). We will return to this point later.

While not being properly a structure, a compliant system is neither a mechanism but something in between, or, better, both at the same time, since it joins some properties (and challenges) of both sorts of mechanical systems. Being closely related to two strong disciplines (machine engineering – or kinematics – and structural mechanics), each of them with its own nomenclature and principles, the science of compliant mechanisms is naturally prone to some sort of ambiguity or vagueness. Let us consider, for instance, the compliant system of Figure 4.14(a). In the view of structural mechanics, this system has an infinite number of degrees of freedom. In the sense of kinematics, the number of degrees of freedom (or *mobility*) is zero (or even negative (Shigley and Uicker, 1980), since the system under consideration responds to any applied load with finite displacements. Of course, the purpose of the system of Figure 4.14(a) is to approximate a rigid-body parallelogram (a mechanism with one degree of freedom) like the one shown in Figure 4.14(b). Searching for an appropriate definition of degrees of freedom which reflects this fact, it is useful to refer to the system's stiffness matrix (which, of course, implies a discrete mathematical model). In the case of a conventional (supported) structure, the stiffness matrix is regular and well conditioned, whereas a rigid-body mechanism (with one fixed link) has a singular stiffness matrix, with rank deficiency equal to the number of degrees of freedom of the mechanism. For a *compliant mechanism*, the stiffness matrix is regular but badly conditioned, i.e. with a certain number of eigenvalues (in the particular case of Figure 4.14(a), just one eigenvalue) which are definitely smaller than the other ones. The number of 'very small' eigenvalues provides the mobility of the compliant mechanism. Of course,

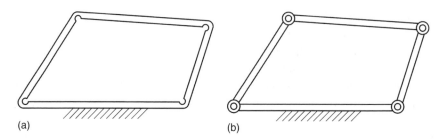

Figure 4.14   A compliant mechanism (a) and the corresponding rigid-body parallelogram (b)

this definition is somehow arbitrary (how small is a very small eigenvalue?), but this kind of arbitrariness is implicit in the nature of compliant systems as mechanical systems with distributed properties. The reader who is interested in the nomenclature-related discussion on compliant systems can refer to Shoup and McLarnan (1971), Her and Midha (1987), Howell and Midha (1995) and Midha *et al.* (1994).

Even more important than nomenclature differences between the disciplines of kinematics and structural mechanics are the diverging principles and criteria which drive priorities in design and optimisation, as will become clear later.

In comparison with conventional, rigid-body mechanisms, compliant systems present several advantages. They have no wear and no backlash, produce less noise, do not release particles, require no lubrication and are suitable for a monolithic manufacture with no need for assembly (Troeger, 1962; Sevak and McLarnan, 1974; Howell and Midha, 1994). Additionally, they possess a better scalability towards small sizes and can reach a very high accuracy level (Bögelsack, 1995; Mankame and Ananthasuresh, 2004; De Bona and Zelenika, 2002).

Most disadvantages of compliant systems are related to the negation of the task separation principle discussed in Section 4.3. Unlike articulated systems with conventional hinges and bearings, compliant systems are fully coupled mechanical systems in which load-carrying capabilities and deformability are shared by all degrees of freedom. The first evident consequence of this fact is a more complex design procedure without specifically 'bearing' and 'yielding' elements. Since compliant systems are characterised by a determined relationship between forces and displacements (a regular stiffness matrix), they possess a load-dependent kinematics: while a given point on the rigid-body mechanism of Figure 4.14(b) moves on a prescribed path independently of the acting loads, this does not hold true for the compliant mechanism shown in Figure 4.14(a). Another disadvantage which is related to the coupled nature of compliant systems is the limited range of motion: a compliant element can only be deformed up to a defined level, beyond which the allowable strain of the material is exceeded. Unlike their rigid-body counterparts, solid-state hinges cannot be designed for an unlimited rotation angle.

For the same reason, the deformation of a compliant system requires a definite amount of mechanical work, which is stored in the compliant components in the form of strain energy. In the literature (Venanzi et al., 2005), this is generally seen as a further disadvantage. Indeed, from the point of view of kinematics – and the large majority of contributions to this topic stems from the scientific community dealing with mechanism design – it is a drawback, since the purpose of a mechanism consists in converting motion and transferring forces as efficiently as possible. While looking at a compliant system as the passive part of a shape-adaptable system, however, one should not forget that providing the system's 'kinematics' – i.e. the large, desired deformation components – with stiffness also involves the capability of carrying a portion of external load in the motion degrees of freedom. A lower portion of those loads has hence to be carried by the actuator, which can – in some cases – be made smaller and lighter despite the need for a higher actuator force. In other words, referring to the mechanical design task as introduced in Section 4.2, the presence of stiffness in the motion degrees of freedom surely moves the problem's activability restriction towards regions of the design space in which the actuator's size is bigger; on the other side, it moves the structural restrictions (stiffness and strength) away from those regions.

Last but not least, compliant systems can be made considerably lighter than pin-jointed mechanisms (Sevak and McLarnan, 1974; Ananthasuresh and Kota, 1995). Of course, this assertion has to be seen in a relative way, since a mechanical system can be made arbitrarily light by reducing its size and hence his load-carrying capability. If properly exploited, regardless, the large optimisation potential of compliant systems can conciliate a high strength-to-weight ratio with a high degree of deformability (more on this later).

The field of compliant systems is attracting growing interest in the scientific and technical communities. In the last few years three monographs have appeared (Smith, 2000; Howell, 2001; Lobontiu, 2003), which offer a thorough, well-documented insight into this fascinating topic. These books collect and systematise a broad range of results, concepts and methods which were elaborated in the last decades of the twentieth century, mostly documented in conference and journal papers.

The topics treated in the literature cover a broad range of aspects, from the analysis and design of single flexible components up to the synthesis of complex systems consisting of rigid and flexible parts. As possible design options for a compliant system, a distinction is usually made between systems with *lumped* compliance, consisting of stiff links and compliant pivots, and systems with *distributed* compliance, in which only flexible elements are employed. Of course, mixed and intermediate designs are conceivable. Among the single components (see Figure 4.15), the choice ranges from simple hinge concepts like leaf-type flexures and notch hinges (Paros and Weisbord, 1965) to more sophisticated architectures like the cross-spring pivot (Eastman, 1937; Haringx, 1949), the cartwheel or Haberland hinge (Haberland, 1981), the split-tub flexure (Goldfarb and Speich, 1999) and the cruciform hinge (Freise, 1949). A thorough overview of different concepts for solid-state hinges is offered in Trease *et al.* (2005). As far as practical applications are concerned, compliant systems are mainly employed in small-scale, high-precision devices, for instance in measurement instruments (Wuest, 1950), micro-assembly tools (Horie and Ikegami, 1995; Hesselbach *et al.*, 1997; Ando *et al.*, 1990) or in stroke amplification devices for

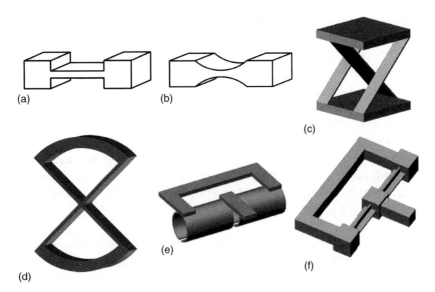

Figure 4.15 Different solid-state hinge architectures: leaf hinge (a), notch hinge (b), cross-spring pivot (c), cartwheel hinge (d), split-tube hinge (e) and cruciform hinge (f). Images (c), (e) and (f) from Canny, Risner and Subramanian, 2002, reproduced by permission of Springer

solid-state actuators and motors (Kota *et al.*, 1999b; Nelli Silva *et al.*, 2000; Tenzer and Ben Mrad, 2004).

A classic issue which has its roots in the early work of Euler (1744) and builds the basis for the modelling of leaf-type hinges and flexible links is the analysis of large deformations in beams. More recently, this subject was addressed by Bisshopp and Drucker (1945), who analysed the problem of an initially straight beam loaded by a force at the free end and provided a closed-form solution involving elliptic integrals. Other significant contributions include the works of Johnson (1937) and Conway (1956) on curved beams of circular shape, a book by Frisch-Fay (1962) and the excellent review by Gorski (1976).

Most of the analytic work done on such nonlinear modelling was of mere academic relevance (Gorski, 1976) for a long time. Among the first scientists to put these results into a practical framework, Burns and Crossley (1968) approached the problem of designing flexible link mechanisms. To this end, they developed a computer-aided procedure in which analytical beam models involving elliptic integrals were used. Closed-form solutions based on elliptic integrals also build the basis of deflection approximations used by Howell and Midha (1995) which allow the description of the motion of a cantilever beam with a small-length flexural pivot located at the fixed end as a rotation around a fixed point (*characteristic pivot*). This kind of approximation constitutes the foundation of one of the most popular techniques for the synthesis of compliant systems: the pseudo-rigid-body model (Howell and Midha, 1994; Howell and Midha, 1996). In the pseudo-rigid-body model, flexural joints are replaced by conventional pivots with a torsional spring as a substitute for the bending stiffness of the solid-state hinge. Of course, this kind of method is well suited to the design of compliant mechanisms with short-length pivots; the accuracy of the pseudo-rigid-body approach decreases as the pivot length increases and becomes impracticable for systems with distributed compliance.

As an alternative to the above kinematics-inspired approach, other authors opted for a more 'structural' view of the synthesis problem. In one of the first works of this kind, Sevak and McLarnan (1974) presented an optimisation-based synthesis procedure which operates on a parameterised nonlinear finite element model. In the application examples reported in the paper, the synthesis procedure operates only on a few design variables which describe shape or size properties of the mechanisms to be designed, which are otherwise known a priori. Possible extensions of the method to more complex synthesis problems, involving a larger number of variables or even including the mechanism topology as unknown, are not discussed. A more versatile method, based on the *homogenisation* method, was proposed by

Ananthasuresh *et al.* (1994). Since it can operate on arbitrary geometries within a prescribed domain, this method is apt to optimise not only the shape of the compliant systems and the size of the single elements, but also the system's topology. The method employs a so-called *cellular microstructure* consisting of cells with parameterised geometry. A further possibility for topology optimisation (Frecker *et al.*, 1997; Kota *et al.*, 1999b) consists of using a lattice structure with fixed nodes and rods of variable thickness (*truss ground structure*). By leaving the material distribution virtually free, topology optimisation techniques possess the charm of being independent of a priori design choices; however, the optimisation results often need a posteriori interpretations and modifications involving the designer's judgement and are still far – at the present state of the art – from being fully systematic. Furthermore, by choosing a topology optimisation approach, the designer indirectly drives the design towards a typical distributed-compliance architecture and penalises a possible lumped-compliance solution (Lu and Kota, 2002).

The problem of diverging, discipline-biased principles and criteria which was mentioned earlier in this section is particularly evident in optimised-based approaches to the design of compliant systems. While lightweight designers typically tend to maximise stiffness (Frecker *et al.*, 1997), the maximisation of compliance is a central concern to kinematics designers (Ananthasuresh *et al.*, 1994), since they naturally tend to keep energy losses in the mechanism – seen as a device with the purpose of transferring motion or force – low.

Trying to conciliate both views inevitably leads to an inconsistent problem formulation ('maximizing the flexibility of a mechanism while simultaneously maximising the stiffness', Kota *et al.*, 1999b). Further, maximising compliance has proven to lead to unfeasible solutions (Kota *et al.*, 1999a). The good news here is that such a conflict is not necessary, since – as already mentioned – the primary requirement of a mechanism is not flexibility, but deformability, in the sense introduced at the beginning of this section. It is therefore likely that future optimisation-based synthesis procedures for compliant systems will explicitly include deformability instead of flexibility constraints.

A further point which deserves particular notice concerns the lightweight potential of compliant systems. Let us look again at Figure 4.14. A conspicuous difference between the two mechanisms appears when considering the load-carrying capability of the hinges as compared with the stiff members: while the rigid-body solution employs hinges with axial stiffness and strength comparable with those of the bars, the compliant solution achieves deformability by substantially reducing the cross-section and therefore the stiffness and strength in the hinge regions. On the other hand, the rigid-body mechanism is affected, as mentioned earlier, by a significant weight increase as compared with a fixed-shape frame built with the same links, whereas the compliant solution leaves the structure weight essentially unchanged (the slight weight reduction due to the cut-outs can be neglected in this context). Now, which concept is more *lightweight*, in the sense given to this term in structural mechanics? This question cannot be answered in a general way, since in both cases the load-to-weight ratio was worsened with respect to a fixed-geometry structure.

Traditionally motivated by small-scale and high-precision applications, in which the advantages of solid-state hinges and flexible elements are decisive (Paros and Weisbord, 1965; Kota *et al.*, 2001), the science of compliant mechanisms evidently did not have enough incentives to deal with this kind of questions up to now. Let us consider the solid-state hinge represented in Figure 4.16(b). Owing to the large reduction in the cross-sectional thickness, the hinge is provided with high deformability in bending about one axis. By properly increasing the cross-sectional width and providing the flexure with opportune transition regions, however, the axial strength of the original prismatic bar can be retained. Comparison with a flexure showing the same thickness reduction but constant cross-sectional area (Figure 4.16(a)) suggests that this kind of high-load hinge is not free from weight penalties, but they are most likely less severe than in a conventional hinge architecture. For the sake of simplicity we do not consider here combined bending deformation and axial load in the hinge; however, the key facts discussed here possess general validity.

While analysing the high-load hinge of Figure 4.16(b) it becomes clear that the volume and, consequently, the weight of the transition regions increase if the thickness of the small-length flexural pivot decreases (all other things being equal), which points to a trade-off between weight penalty effects and hinge bending deformability. Such a trade-off is absent in a conventional hinge technique (where, so to say, full deformability is charged to the designer – in terms of additional weight – even if he needs a limited swing angle) and

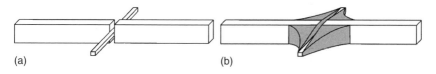

(a)   (b)

Figure 4.16   Small-length pivot with constant cross-sectional area (a) and high-load flexural hinge (b)

is surely of central relevance in the lightweight optimisation of compliant systems.

Experimental tests conducted on hinges with linear width change (Pasini, 2004) show a substantially two-dimensional strain distribution for the bending load case. As a consequence, the deformability (maximum swing angle) of this kind of hinge is essentially a function of its longitudinal cross-sectional geometry – as well as of the allowable strain of the material – and can therefore be studied on the basis of a simple leaf-type hinge model. For a pure moment loading, the maximum swing angle of a leaf-type hinge is given by

$$\Theta = \frac{2\varepsilon_{\text{all}}l}{t}. \tag{4.3}$$

where ε_{all}, l and t represent the allowable strain, the hinge length and the hinge thickness, respectively (see Figure 4.17(a)). Note that this expression is valid also for large deflections. According to (4.3), it is possible to increment the hinge thickness while retaining the same degree of deformability if the hinge length is increased by the same factor (see Figure 4.17(b)). Referring again to Figure 4.16(b) and to the mentioned weight penalty effects, it is hence evident that they can be arbitrarily reduced by choosing a larger pivot length, since this allows for a reduction in the required width of the flexure. The buckling behaviour of the hinge and the shear bending resistance remain virtually unchanged when reshaping the hinge in this way.

The trade-off between short and large pivot length is a stiffness one: by reducing stress concentrations, a large pivot length increases the strength-to-weight ratio but leads to undesired deformations under load (*axis drift* or *parasitic motion*, see Trease *et al.*, 2005). In the specific case examined above, the assumption of a pure moment loading loses validity as pivot length increases, and the mutual motion of the hinge ends becomes increasingly load dependent. By ably combining more than one long leaf-type spring, however,

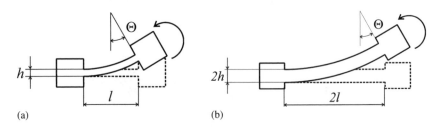

(a)   (b)

Figure 4.17 A small-length (a) and a large-length (b) flexure with the same deformability

these drawbacks can be considerably reduced without losing the advantages of low stress concentrations. A prominent example is the already-mentioned cross-spring pivot consisting – in its simplest form – of two leaf springs arranged as shown in Figure 4.15(c). It shows an almost load-independent kinematics which closely resembles the one of a pin-jointed hinge. A variation of the cross-spring concept is implemented in the Haberland or cartwheel hinge (Haberland, 1981): by connecting the two leaf springs together at the intersection (Figure 4.15(d)), the hinge obtains a prismatic shape which offers clear manufacturing advantages.

In some sense, the cartwheel hinge represents a precursor of planar compliant systems with distributed compliance, which has been the object of extensive research work in the last few years (Kota, 2001; Kota *et al.*, 2001; Hetrick and Kota, 1999; Frecker *et al.*, 1997; Yin and Ananthasuresh, 2003).

4.5.2 Compliant Systems for Airfoil Shape Adaptation

In addition to the advantages discussed in the previous subsection, compliant-systems-based solutions to airfoil shape control (especially the ones relying on distributed compliance) profit from being inherently smooth in the deformation pattern and therefore aerodynamically favourable. A large part of the early approaches to enhanced geometry control in airfoils makes use of flexible elements (skin or ribs) which shape the upper surface (Rocheville, 1932) or the whole outer surface of the airfoil (Holle and Judge, 1916; Antoni, 1928; Grant, 1935; Lyon, 1965). Later concepts, which were conceived after the box-girder construction had become customary, focused on local deformations of the nose and trailing-edge regions (Pierce, 1978; Lee and Hanlon, 1977; Piening and Monner, 2000; McKinney, 1977; Ash and Walley, 1979; Rowarth, 1981; Statkus, 1982), and include the use of flexible skins as well. These concepts can, however, be considered only marginally compliant, since the skins do not accomplish a substantial load-carrying task and can be essentially seen as a smoothening cover for the underlying rigid-body kinematics. Additionally, in those cases in which the skin covers the entire deformable region of the airfoil, the skin contour needs to be opened to avoid exceeding strain limits, and linear slides are introduced at the interruption. This, again, contradicts the compliance principle (structural strain instead of finite motion through sliding parts) and – apart from aerodynamic disadvantages – further reduces the skin's load-carrying contribution.

A fully compliant approach to shape-adaptable airfoils (the *belt-rib* structural concept) was devised by the author in the second half of the 1990s and constitutes the subject of the following subsection. In the framework of the 'Smart Wing' project (Bartley-Cho *et al.*, 2004), a trailing-edge control surface was conceived which allows large and fast deflection of the trailing-edge region ($\pm 25°$ at 3 Hz). It employs a closed skin built of a highly extensible material (silicon) in order to allow the large membrane strains occurring during deflection. For out-of plane loading (i.e. in the direction of aerodynamic load) the silicone skin is supported by a honeycomb core which is compliant to the skin's extensional deformation. Finally, the structure is stiffened by a composite leaf spring placed in the symmetry plane.

One of the peculiar characteristics of shape control of airfoils is their multiple output nature. Unlike the typical single-output applications (crimping devices or stroke amplification devices for actuators) for which the systematic, optimisation-based design methodologies mentioned in the last subsection were developed, airfoils are loaded by distributed external loads and have to fulfil deformation and stiffness requirements in a large – theoretically infinite – number of degrees of freedom. A synthesis approach which explicitly deals with a multi-output design problem was presented by Saggere and Kota (1999) and applied to the shape control of an airfoil in its leading- and trailing-edge regions. The central idea of the method is to use an optimised compliant truss to transfer the force of a single actuator to a set of so-called active points on a flexible surface. Kinematics synthesis aspects and maximisation of compliance in the active deformation mode are primary concerns in this approach: the skin's bending stiffness is set to its minimum value (as prescribed by manufacturing constraints) and a least-squares measure of the deviation between desired and actual deformation is chosen as the cost function (Lu and Kota, 2003). Prior to the synthesis of the compliant mechanism, a minimum number of active points are determined according to geometrical criteria. The point locations also define the basis grid for the truss optimisation.

Another study (Frank *et al.*, 2004) deals with compliance-based concepts for global airfoil shape modification (i.e. not limited to the leading- or trailing-edge regions). Mechanisms with lumped as well as with distributed compliance are considered as separate design options. In both cases, the mechanisms are supported by an internal box structure which is in charge of carrying most of the bending and torsion loads of the wing structure. The solutions which are based on lumped mechanisms use a segmented chain to define the wing profile and a set of internal links apt to reduce the mechanism's mobility to one, in order to make the wing geometry controllable with a single actuator. Minimisation of internal stresses in links and joints as well as a restriction on the actuator force are used as selection criteria for the mechanism's configuration. The synthesis procedure for distributed compliance implements the energy efficiency maximisation approach initially proposed by Ananthasuresh *et al.* (1994), modified for multi-output mechanisms. The choice of possible target shape changes is limited to contour deformations with no longitudinal strain, a feature shared with belt-rib airfoils (see below).

Use of compliant truss mechanisms with lumped compliance to produce large deformations in wing structures was proposed by Ramrakhyani *et al.* (2005). The mechanisms are provided with a mobility greater than zero (in the sense defined at the beginning of the previous subsection) and controlled by tendon actuators. Owing to the positive mobility, the structure can be deformed without storing a large amount of strain energy in the passive members. A compliant skin able to sustain large membrane strains covers the compliant active truss. Two synthesis options were addressed by the authors: topology optimisation and systematic arrangement of standard cells.

Spadoni and Ruzzene (2005) propose the use of chiral compliant mechanisms as the structural core for shape-changeable airfoils. They investigated different design options for the chiral geometry and assessed the good configurability properties of this kind of design, which enables a wide range of compliance requirements to be matched.

4.5.3 The Belt-Rib Airfoil Structure

The basic idea of the belt-rib structural concept for airfoils (Campanile and Sachau, 2000) is the replacement of the plate-fashioned rib in the conventional box-girder wing construction with an optimised compliant planar mechanism (*belt rib*). The primary purpose is an active modification of the airfoil section; other modes of deformation – like twisting or bending – can be attained as a secondary effect by exploiting structural or aeroelastic coupling. The airfoil is provided with a closed skin of conventional kind which is bonded to the outer surface of the belt ribs. The belt-rib construction preferably does not include spars; the nose and trailing-edge skin panels are used as shear webs instead. Alternative designs with a single rear spar and/or a compliant fore spar are possible options. A belt-rib airfoil segment (a) is exemplarily shown in Figure 4.18, together with the corresponding set of belt ribs (b). The belt-rib airfoil structure is conceived to retain most of the structural qualities of a conventional, fixed-shape wing construction while allowing a substantial degree of deformability. It is devised as a continuous, fully compliance-based, shape-adaptable system without sliding parts.

In its basic layout, the belt-rib airfoil is a compliant system of mixed kind. The skin and the outer shell of the ribs (*belt*) are provided with distributed compliance; the *spokes*, which are responsible for stiffening the ribs in their plane, are designed as rigid elements; finally the connections between spokes

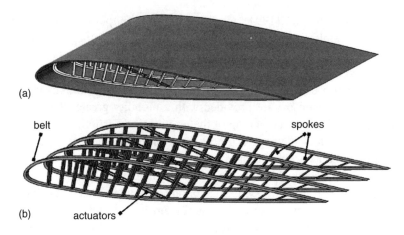

(a)

belt   spokes

(b)   actuators

Figure 4.18   A segment of a belt-rib airfoil (a) with the corresponding set of belt ribs (b)

and belt are realized as flexural hinges (lumped compliance). Alternative design options with entirely distributed compliance are conceivable as well.

The spokes' configuration is optimised to be compliant to a previously chosen *deformation mode*, which somehow represents the kinematics of the belt rib as a mechanism. The deformation mode defines the target shape change of the airfoil section but for a single parameter, which can be controlled by a proper actuator. For deflections which are to conform to the deformation mode the belt rib shows virtually the same stiffness as the sole belt, whereas the stiffness related to other deformation components is comparable with that of a conventional plate rib. The belt rib's kinematics is freely configurable; the only limitation which applies to the choice of the deformation mode is the inextensibility condition: to avoid a membrane stress state during activation, only deflection patterns with constant arc length are allowed. The optimisation of the spokes' layout is performed on the basis of a modal procedure specially developed for this purpose (Campanile *et al.*, 2004b).

The restriction of the belt-rib's kinematics to virtually one mode (or to a small number of modes, if the above-mentioned layout optimisation is extended to more than one deformation mode) allows a broad choice of options for actuation. Linear actuators obliquely connecting one point on the upper side of the belt with one point on its lower side (see Figure 4.18) represent perhaps the simplest option. If externally activated tendons (e.g. Bowden cables) or shape-memory alloys are used, a truss configuration allows working only with tension actuator forces. Alternatively, active material can be embedded into the belt, which can in this way be directly activated in a bending mode.

A peculiar characteristic of the belt-rib design philosophy is the absence of a general claim for maximum compliance; on the contrary, stiffness is intentionally kept over a prescribed bound, usually by acting on the belt design. By choosing a higher bending stiffness of the belt, the system's structural redundancy can be enhanced and the forces acting on the actuator in the presence of external loads decreased. Furthermore, a stiffer belt guarantees contour smoothness and reduces undesired deflections under load.

The belt-rib design was applied to several practical cases, among others to the swept outer wing of a small high-performance aircraft and to a landing flap of a large passenger aircraft. Numerical calculations as well as experimental tests on different prototypes confirmed the feasibility of the concept in a realistic load environment. In the case of the high-performance aircraft, the belt-rib airfoil (1.8 m span length, 1.0 m chord) allows changes in section camber ratio of up to 4 percentage points and can withstand a load of over 91 000 N without reaching a critical stress state (Kiehl, 2006).

A fully compliance-based design offers attractive manufacturing options, which reduce or even eliminate the need for assembly. An integrally manufactured composite belt-rib prototype is shown in Figure 4.19. A hybrid carbon/glass fibre system is used as reinforcing material, with the glass fibre component mainly used in the bending region of the flexures. By means of a novel sewing-based technique the solid-state hinges can be firmly secured to the belt fibres in a dry state (see Figure 4.20(b)), which provides them with proper tensile strength. The photo in Figure 4.20(a) shows a spoke with the corresponding hinges after curing. All structural elements of the belt rib are simultaneously injected by resin transfer moulding. The prismatic shape of the belt rib allows for an efficient manufacturing procedure in which a wider block is moulded and cured first and then cut into several single ribs of the desired width.

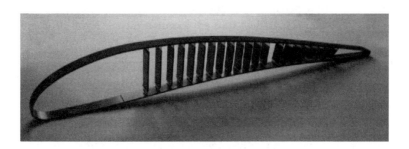

Figure 4.19   Belt-rib prototype

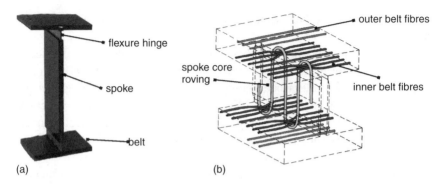

Figure 4.20   Belt-rib section segment (a) with manufacturing details (b)

The belt rib of Figure 4.19 was realised for a 1:2 scale prototype of an Airbus A330/340 landing flap. For details on deformability, loads and weight the reader can refer to Campanile (2006).

4.6   EXTENDING COUPLING TO THE ACTUATOR SYSTEM: COMPLIANT ACTIVE SYSTEMS

4.6.1   The Need for a Coupled Approach

From the considerations of the preceding section it should be clear that the load dependence of kinematics plays an important role – together with weight, strength and deformability – in the multiple trade-off which rules the design of compliant systems. On the one hand, it can be virtually eliminated, at the price of a poor strength-to-weight ratio, by using lumped compliance systems with small-length flexures and otherwise rigid links; on the other hand, distributing compliance allows reduced stress concentrations while keeping the weight penalty at a low level, but makes kinematics increasingly load dependent. It is therefore evident that the potential of compliant systems for lightweight shape-adaptable systems can be exploited at best in the framework of a coupled design which includes the actuator system.

In some cases, when particular restrictions apply, it is not possible to deal with the stiffness and deformability restrictions in a largely uncorrelated way, as in the case of the high-load flexure of Figure 4.16(b). Then the simultaneous fulfillment of both requirements at the level of the passive compliant system can become problematic or even impossible, and the stiffness contribution of the actuator turns out to be essential. In such cases, coupled analysis and synthesis of compliant system and actuator is necessary regardless of the load dependence of kinematics.

The following example is taken from a project which was carried out at the German Aerospace Centre (DLR) in the late 1990s. A shape-adaptive wind-tunnel model had to be designed and realised within the project (Campanile *et al.*, 1999). The required geometry changes included a so-called transonic bump and a hingeless trailing-edge flap. Even though lightweight optimisation was not a concern for the wind-tunnel model, compliant systems were chosen for both measures owing to their clear advantages at small scale.

The concept which was adopted for the hingeless flap was based on a distributed-compliance layout (see Figure 4.21). This design responded very well to manufacturing and aerodynamic requirements as well as to the restrictive space constraints. Shape-memory alloy wires were chosen for actuation. A minimum swing angle was prescribed for the active flap as well as the maximum allowed deviation from the target contour under load. Destined to operate in two-dimensional flows, the wind-tunnel model was additionally required to have constant deformation behaviour in the span direction. This called for a prismatic geometry of the compliant system and virtually reduced the design variables to those defining the thickness and the material of one single leaf spring. The spring length was set to the largest value compatible with the space constraints: beyond this limit the airfoil thickness at the spring end would have become too small to accommodate the attachment between actuator wires and structure without causing manufacturing problems and leading to unacceptably low values of the actuator lever arm. In the considered case, therefore, the deformability versus stiffness trade-off is essentially ruled by just one design variable, namely the spring thickness (some influence is exerted by the material choice, but within narrow limits).

In this case, deformability and stiffness are indeed directly conflicting qualities: ensuring proper deformability means setting an upper bound for the spring thickness, while the stiffness requirement implies a lower bound for it. It can therefore happen that the design problem admits no solution if

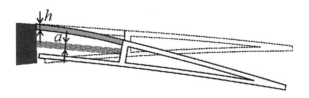

Figure 4.21   Design of a hingeless trailing-edge flap

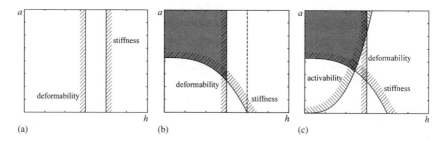

Figure 4.22   Design space graphs for the hingeless trailing-edge flap of Figure 4.21

limited to the passive compliant system. If the stiffness contribution of the actuator is taken into account, however, any required level of overall stiffness can be reached – without affecting deformability – by properly choosing the cross section of the wire bundle.

This effect is exemplarily illustrated by the design domain graphs of Figure 4.22. The graphs are traced on the h-d–plane, with h as the leaf-spring width and a as the cross-section of the actuator wire bundle per unit width in span direction. If the actuator's stiffness contribution is not considered (graph (a)), the actuator cross-section does not come into play and the feasible design domain, identified by the restrictions $h \leq h_d$ and $h \geq h_s$, is an empty set (no feasible solutions) if $h_d < h_s$. The contribution of the actuator material to the overall stiffness modifies the corresponding restriction boundary to a curve of the kind $h^3 + Ba = h_s^3$, with $B > 0$, which always intersects the boundary of the deformation restriction, providing a finite region of feasible designs (graph (b)). A possible reduction of the feasible design domain (graph (c)) can result from the activability restriction, which requires the actuator wire bundle to provide enough active force to bend the leaf spring by the required amount and takes the form $a/h^3 \geq C > 0$.

4.6.2   Solid-State Actuation for Solid-State Deformability

Solid-state actuation provided by active materials like shape-memory alloys represents the ideal complement to the solid-state deformability principle implemented in compliant systems for a multitude of reasons, thus providing a highly promising option for lightweight shape adaptation.

A first consideration in this sense is related to the attitude of active materials to provide load-carrying capabilities in multiple directions. In order to activate a mechanism of the conventional kind with, say, one degree of freedom, an actuator system is needed which is able to control this single degree of freedom and is not required to provide a structural contribution in other directions. A conventional actuator (e.g. a spindle linear actuator or a hydraulic cylinder) is built to be loaded in one single mode and therefore fits well with a conventional mechanism technique. A solid-state actuator fundamentally interacts with the structure by means of distributed stresses and strains and therefore offers the possibility of carrying loads in multiple directions. This, in turn, helps making the passive component lighter and possibly more deformable. Exploiting this feature, however, implies a high degree of integration between active and passive material. Most of the documented applications of solid-state actuators for shape adaptation (like the above-described hingeless flap application or when, for instance, piezoelectric actuators are used as single-point devices) are still kept at a relatively low integration level, which virtually restricts the load-carrying contribution of the active material to the degrees of freedom of active deformation.

But even when restricted to the activation direction, solid-state actuators offer superior load-carrying capabilities due to the strain-induced actuation principle. Unlike most conventional actuators which primarily generate stresses or forces, solid-state actuators work by producing displacements and strains; forces and stresses are secondarily generated when the active strain is hindered by external loading. This allows solid-state actuators to supply stiffness in the actuation direction in a very efficient and reliable way. Let us consider again the above hingeless flap example. If the shape-memory wires were replaced by an electromagnetic linear actuator of the solenoid type, the stiffness contribution provided by the actuator, given in general by

$$k_A = \frac{dF}{du},\tag{4.4}$$

would disappear as long as the actuator would be seen as an open-loop device with a constant active force. A requirement conflict of the type described in Section 4.6.1 could not be solved in this case. In Equation (4.4), the symbols F and u denote the actuator force and stroke, respectively. The derivative on the right side depends in general on the actuator's characteristic (a family of curves in the F–u–plane, parametrically dependent on the actuator drive input) as well as on the way the actuator is controlled. Considering now the case of constant driving voltage or current, some interdependence between actuator force and stroke will appear and the actuator stiffness will become appreciable. This sort of virtual stiffness can be made arbitrarily high by means of a proper feedback control (it reaches a theoretically infinite value if the actuator stroke is kept constant by the controller), but it relies completely on the actuator's energy supply as well as on the control hardware. As a consequence of limited driving dynamics, it can undergo bandwidth limitations which can cause problems if the stiffness restriction results from dynamic constraints (e.g. flutter prevention).

While being affected as well, to some extent, by the control law or, more generally, by the way the actuator's input quantity is allowed to change (Campanile et al., 2004a), the stiffness provided by solid-state actuators is primarily passive, i.e. based on material loading, and thus not only always available (regardless of energy supply, controller functionality and dynamics), but also comparatively high.

Strain-induced actuation can be successfully emulated by actuators based on conventional principles by means of blocking or self-locking devices (for instance, in electromechanical spindle actuators) but usually at the price of increased complexity, additional weight and/or reduced efficiency.

A further advantage of active materials consists of allowing for distributed actuation. Avoiding load concentrations helps to keep weight low and becomes of essential importance for compliant systems with distributed compliance, in which point forces can cause overload of the flexible members. The good scalability of solid-state actuators can also play an important role when distributed actuation is to be approximated by a large number of single small-size actuators. The above hingeless flap design, for instance, includes 105 single wire bundles distributed over a length of 1 m in the span direction (Campanile et al., 1999), while the contour bump area in the same wind-tunnel model incorporates 84 hybrid solid-state actuators (essentially consisting of a composite spring activated by shape-memory wires) distributed on a surface of about $0.08\,m^2$ (Campanile et al., 2004a).

Active materials represent an attractive option for lightweight shape adaptation, not least because of their high specific performance (Kuribayashi, 1993; Jänker and Martin, 1993; Huber et al., 1997; Giurgiutu et al., 2000; Carpi et al., 2005). In particular, shape-memory alloys and electro-active polymers (EAPs) show high values of output energy density (mechanical work performed by the actuator in a single activation cycle related to the actuator's volume or weight); piezoelectric materials offer an excellent dynamic performance, expressed by the power density (maximum specific deliverable power for periodic activation), and constitute a very promising choice for airfoil shape adaptation when aeroelastic amplification effects are exploited (see next section).

Further, solid-state actuation can be modelled at the material level, which leads to an easier analysis of actuator performance as a function of its geometry and offers clear advantages while designing and optimising the compliant structure and the actuator system in coupled form. Model-based performance analysis of conventional actuators presents a high degree of complexity, so that their design relies widely on statistical methods (Huber *et al.*, 1997). Finally, most of the general advantages of compliant systems which are related to their solid-state nature apply to solid-state actuation as well: absence of wear, backlash and particle release, no need for lubrication and easier manufacturing.

4.6.3 Challenges and Trends of Structure–Actuator Integration

Most of the advantages reported in the previous subsection increase substantially with the degree of integration between structure and actuator. Efficient procedures for coupled analysis and optimisation as well as suitable manufacturing techniques for actuator embedding are therefore of fundamental importance to the successful development of solid-state actuated compliant systems for lightweight shape adaptation.

In the last few years, a broad range of methods and criteria for the optimisation of active structures have been developed. However, these are mostly conceived for response control applications rather than for geometry adaptation and mainly focus on optimising the actuator system for a given structure; coupled approaches dealing with the simultaneous design of structure and actuator are comparatively rare. A large part of the contributions present in the literature deals with optimal actuator placement for a predetermined structure, mostly to the aim of vibration control. Well-documented reviews on this topic can be found in Padula and Kincaid (1999) and Frecker (2003). In the context of shape adaptation, the work by Strelec *et al.* (2003) is worth mentioning, in which the shape-memory actuator system of a reconfigurable airfoil is optimised by means of genetic algorithms. However, the design of the passive structure is considered as given also in this study.

Conversely, in several works on solid-state actuated compliant systems, optimisation procedures deal with the sole design of the passive component while taking the actuator properties as given. An example in this sense is given by the already-mentioned work of Nelli Silva *et al.* (2000) on the design of flextensional actuators. Other references of a similar kind are contained in Frecker (2003).

Simultaneous optimisation of both structure and actuator for an active plate was performed by Chattopadhyay *et al.* (1999). No shape adaptation, however, is pursued in this work; the main objective is dynamic response control, and no constraints requiring minimum deformability or active deflection are included in the optimisation problem. A coupled design procedure for the contour bump region of the above-mentioned shape-adaptive wind-tunnel model was presented in Campanile *et al.* (2004a). The procedure is based on analytical, numerical and experimental modelling of the compliant system's as well as of the shape-memory wires' behaviour and does not make use of formal optimisation techniques. Design of a flexural actuator built as a sandwich structure with shape-memory alloy faces and a corrugated core was undertakes by Lu *et al.* (2001) by means of formal optimisation. Yielding and buckling constraints, a power consumption threshold and a lower limit for the active deflection were inserted into the problem's restrictions.

A fundamental prerequisite for the coupled design of solid-state actuator and structure is the availability of proper simulation tools. A large amount of work has been performed in the last few decades with the purpose of describing the complex behaviour of active materials and combining the resulting mathematical models with structural analysis techniques, like the finite element method. For a review on this topic, the reader can refer to Chopra (2002).

As far as layout options for embedded solid-state actuations are concerned, several alternatives have been investigated in the past few years. First studies on the integration of piezoceramic thin plates date back to the 1980s (Crawley and de Luis, 1987). Encapsulation techniques aiming at producing easy-to-handle hybrid modules including piezoceramic plates, electrodes and contacting circuitry were devised in the following years (Lazarus *et al.*, 2002; Wierach, 2003) and implemented in commercial products (Wierach and Schönecker, 2005). An essential contribution to the use of piezoceramic fibres as embedded actuator elements was provided by Bent and Hagood (1997) who devised a special electrode configuration (*interdigitated electrodes*) to provide the piezoelectric fibres with the high electric field necessary for effective activation. Fibre-like elements with rectangular cross-section, which benefits from better contacting conditions and efficient manufacturing, were subsequently developed by NASA (Wilkie *et al.*, 2000; Wilkie *et al.*, 2003) and successfully marketed in the form of hybrid modules with interdigitated electrodes and polymer encasing (Wierach and Schönecker, 2005). More recently, Pini and Ermanni (2004) proposed the use of electric conductive polymers as a substitute for interdigitated electrodes in encapsulated fibre actuators. Further promising architectures include fibre-reinforced shape-memory composites and hybrid components based on shape-memory thin films (Wei *et al.* 1998; Winzek *et al.*, 2004; Fu *et al.*, 1994).

4.7 A POWERFUL DISTRIBUTED ACTUATOR: AERODYNAMICS

4.7.1 The Actuator Energy Balance

The actuator system of a shape-adaptable structure must provide enough force and stroke to produce the desired shape changes (activability restriction). This implies equating the mechanical work needed to produce the required deformations to the mechanical work which can be supplied by the actuator. Energy balance considerations are particularly useful in the framework of a coupled design procedure since they are only weakly dependent on design variables like actuator design and geometry. On the demand side, actuator work essentially depends only on the required shape changes and the stiffness of the passive system; on the supply side, it directly relates to the overall actuator volume and weight. Keeping the required actuator work low is therefore a primary concern, especially with regard to lightweight optimisation.

In the case of airfoil shape adaptation, the actuator energy balance can be substantially improved by extracting energy from the air stream. Two sorts of measures can be conceived in this sense, depending on the kind of relationship between aerodynamic forces and airfoil displacements. If constant aerodynamic forces are concerned, mechanical work performed by the flow on the structure in a particular region of the airfoil surface can be transferred to another region and used there to equilibrate work of the opposite sign (i.e. performed *by* the structure *on* the flow). Linear aerodynamic forces, on the other hand, involve a quadratic contribution to the energy balance which, if of opportune sign, can partially compensate for the strain-related component of the actuator work demand (aeroelastic servo effects). Further terms can be considered if nonlinearities of an aerodynamic or structural kind are involved.

Finally, air flow energy can be used to deform an airfoil without employing any actuators, if the airfoil structure is provided with selectable stiffness (semi-active aeroelastic airfoils).

4.7.2 Balancing Kinematics by Partially Recovering Energy from the Flow

The idea of balancing kinematics in airfoil shape adaptation by inducing work of opposite sign in different regions of the airfoil is quite old; as reported in Section 4.4, it was implemented as early as 1915 in the variable-camber design devised by Holle and Judge (1916). Partially balanced control surfaces of a conventional kind were used in World War I fighters (Loewy, 1997). Variable-camber airfoils built according to the belt-rib design also profit from a partially self-balanced kinematics: for the case of the Airbus A330/340 landing flap, the required actuator work could be reduced by more than 50 % as compared with an equivalent airfoil with a conventional flap (Campanile and Anders, 2004).

4.7.3 Active and Semi-Active Aeroelasticity

The idea behind active aeroelasticity is to modify the flow conditions by actively generating proper airfoil shape changes and inducing in such a way further deformations as a result of the aerodynamic load. In its simplest form, active aeroelasticity exploits aeroelastic servo effects by controlling the static fluid–structure interaction: negative aerodynamic stiffness weakens the structure and allows for large deformations at a comparably low cost in terms of actuator work. The amplification factor (strain energy related to actuator work) can be made arbitrarily high by approaching static aeroelastic instabilities, and airfoil shape adaptation can be achieved at virtually vanishing actuator work by rendering the structure aeroelastically unstable. Artificial stabilisation can be achieved by feedback control, as proposed by Loewy and Tseng (1993), but this option makes structural integrity – and possibly safety of the whole aircraft – depend on the reliability of the control system. A safer alternative can be conceived on the basis of solid-state actuation: an aeroelastically unstable compliant system (which can therefore be deformed by the actuator at virtually no effort) can be rendered passively stable by bringing the actuator stiffness into play (Campanile and Anders, 2004).

The idea of exploiting aeroelastic servo effects inspired the Active Flexible Wing program (Perry et al., 1995) as well as the follow-up research program named Active Aeroelastic Wing (Pendleton et al., 2000). The target of these large projects was to investigate the possibility of exploiting aeroelastically induced wing torsion deformations for enhanced roll performance of a fighter aircraft. Control of wing deformation was achieved by means of conventional flaps. Two-dimensional aeroelastic servo effects for belt-rib airfoils were addressed in Campanile and Anders (2004). A special thin-walled beam model with section deformations was introduced in Campanile and Mierheim (2005) as a basis for a three-dimensional aeroelastic analysis able to handle combined modes with bending, torsion and camber change components.

Another form of active aeroelasticity involves non-stationary aerodynamic effects to obtain a periodic airfoil deformation with desired phase lags

between different motion components (e.g. with the target of thrust generation) without the need for complex multi-channel actuation. Such an expe-
dient plays an essential role in biological flight: the asymmetrical layout of
feathers in birds, for instance, generates favourable changes of the angle of
attack during periodic wing flapping (Lindhe Norberg, 2002). Engineering
solutions based on this kind of active aeroelasticity can be found in the wing
design of micro air vehicles (Ifju *et al.*, 2006).

Concerning semi-active aeroelasticity, the concept presented by
Amprikidis and Cooper (2004) is worth mentioning: moveable spars are used
to modify the airfoil's stiffness properties, inducing in this way controllable
torsion and bending deformations at the expenses of the flow energy. An
attractive option involving active and semi-active aeroelasticity consists of
using modifiable stiffness to adjust the occurrence of aeroelastic instabilities
to different flight conditions (in particular, different flight speeds).

4.8   THE COMMON DENOMINATOR: MECHANICAL COUPLING

A common principle subtends the research trends addressed in the previous
three sections: the progressive dissolution of the task separation paradigm.
Components which were clearly separate in a conventional design approach
become increasingly interlocked and coupling effects which were carefully
avoided are now deliberately induced. Compliant systems can be seen as
the fusion of mechanism and structure, and active compliant systems with
embedded actuation include integration between mechanism and actuator.
Active and semi-active aeroelasticity represents coupling between structure
and flow, while aerodynamically balanced kinematics reflects the interaction between mechanism and flow. These 'bulkhead' removals are, to some
extent, mutually interdependent: the inherent load dependence of kinematics
in compliant systems as well as the distributed nature of compliance-based
deformability call for integrated, distributed actuation, while higher flexibility requires structure–flow coupling to be included at full title in the design
process.

A coupled view of mechanical synthesis modifies not only the way in
which the design problem is solved, but to some extent the problem definition
itself. In airfoil design, for instance, stiffness requirements usually have an
aeroelastic background; if aeroelasticity is part of the design problem, the
need to prescribe a lower stiffness limit disappears. Flexibility does not
constitute a problem as long as loads are safely carried and the resultants of
the aerodynamic forces – which finally determine the aircraft's motion – can

be controlled. But even if this can render the design problem easier to define or to describe, mechanical design in coupled form is incomparably more complex to handle.

The other side of the coin is represented by clear performance advantages which are the natural result of a broader design space, for instance higher configurability and lower weight, as well as the disappearance of typical interface problems like wear and backlash.

4.9 CONCLUDING REMARKS

There is a series of interrelated trends of research which points at a paradigm change in aeronautical design or – more generally – in the synthesis of mechanical systems. Established design principles, concepts and devices which played a fundamental role in technological development are now seen in a more critical way owing to the perspectives opened by progress in computer-assisted modelling and optimisation, manufacturing technology and materials science. Classic kinematics based on pin-jointed hinges and point actuation by single-stroke devices are making room – especially in weight-sensitive applications – for compliant systems and distributed, solid-state actuation. Aeroelastic phenomena are no longer seen as deleterious effects but are more and more considered as a source of performance enhancement. Synergetic integration of all these aspects will lead to a new class of mechanical systems able to join geometric configurability, strength and light weight in an unprecedented way. As far as aeronautical applications are concerned, such lightweight shape-adaptable systems are likely to pave the way for the realisation of an old dream of humankind: bio-mimetic flight.

Biological suggestions can be implemented at very different levels, ranging from a simple hingeless flap up to a large-scale ornithopter, in which not only flight controls but even thrust production rely on smooth airfoil deformations; up to which level, for which applications and on which time scale bio-mimetic solutions will become feasible cannot be easily answered at the present time – what it is certain is that in any case the availability of a technology for lightweight shape adaptation will play a crucial role in this context.

The key aspect of the above paradigm change is the progressive dissolution of the task separation principle: the challenges of a fully coupled design are accepted with the prospect of an unrestricted optimisation potential. The primary concern of future research efforts will therefore be directed towards the development of efficient and reliable procedures for coupled design, which will in all probability widely rely on formal optimisation techniques but will surely take advantage of the systematic development of special elements like high-load, low-weight solid-state hinges as well. Coupled analysis tool for multi-field problems (flow–structure or actuator–structure interaction) will be a central issue in this framework, as well as modelling of complex material behaviour and geometric nonlinearities. Further progress in manufacturing and integration technologies as well as in materials science is also fundamental. Besides the development of active materials, special focus is likely to be given to materials combining high allowable strain with good strength properties, like for instance superelastic shape-memory alloys.

Of course, the coupling nature of lightweight shape-adaptable systems involves several other disciplines and issues beyond applied mechanics (e.g. sensor technology, signal processing and control) which could not be addressed in the context of this chapter. While not disputing the importance of these components, the author believes that mechanical aspects deserve particular attention owing to the close relationship between mechanical qualities and overall system weight.

ACKNOWLEDGEMENTS

This work is dedicated to the memory of Paolo Santini, a great scientist and teacher, who passed away while this manuscript was being prepared. Most of the author's enthusiasm for aeroelasticity and lightweight design has its roots in his genius and charm.

REFERENCES

Amprikidis, M. and Cooper, J.E., 2004. 'Experimental validation of wing twist control using adaptive internal structures', AIAA Paper No. 2004–1884.

Ananthasuresh, G.K. and Kota, S., 1995. 'Designing compliant mechanisms', *ASME Mechanical Engineering*, November, 93–96.

Ananthasuresh, G.K., Kota, S. and Gianchandani, Y., 1994. 'A methodical approach to the design of compliant micromechanisms', Solid-State Sensor and Actuator Workshop, Hilton Head, SC, USA, 189–192.

Ando, Y., Sawada, H., Okazaki, Y., Ishikawa, Y., Kitahara, T., Tatsue, Y. and Furuta, K., 1990. 'Development of micro grippers', *Microsystem Technologies*, 90, 844–850.

Antoni, U., 1928. 'Flugzeugtragfläche mit veränderbarer Wölbung und auf elastischen Rippen ruhender Bespannung', German Patent No. 464373.

Ash, G.J. and Walley, G.D., 1979. 'Variable-camber aircraft wings', UK Patent Application No. 2 006 133 A.

Bartley-Cho, J.D., Wang, D.P., Martin, C.A., Kudva, J.N. and West, M.N., 2004. 'Development of high-rate, adaptive trailing edge control surface for the Smart

Wing Phase 2 wind tunnel model', *Journal of Intelligent Material Systems and Structures*, 15(4), 279–291.

Bekker, M.G., 1956. Theory of land locomotion, University of Michigan Press, Ann Arbor, 383.

Bent, A. and Hagood, N.W., 1997. 'Piezoelectric fiber composites with interdigitated electrodes', *Journal of Intelligent Material Systems and Structures*, 8(11), 903–919.

Berry, P., 2000. 'Segmented flap with variable camber for aircraft wing', US Patent No. 6,123,297.

Bisshopp, K.E. and Drucker, D.C., 1945. 'Large deflections of cantilever beams', *Quarterly Journal of Applied Mathematics*, 3(3), 272–275.

Bögelsack, G., 1995. 'Nachgiebige Mechanismen in miniaturisierten Bewegungssystemen', Proc. 9th World Congress on the theory of machines and mechanisms, Milan, Italy, 3101–3104.

Buchanan, R.A., 2002. Brunel. The life and times of Isambard Kingdom Brunel, Hambledon, London.

Burns, R.H. and Crossley, F.R.E., 1968. 'Kinetostatic synthesis of flexible link mechanisms', ASME Paper No. 68-Mech-36.

Campanile, L.F., 2004. 'Weight optimisation of hinges for light mechanisms: criteria and design aspects', *Journal of Structural and Multidisciplinary Optimization*, 28, 206–213.

Campanile, L.F., 2005. 'Initial thoughts on weight penalty effects in shape-adaptable systems', *Journal of Intelligent Material Systems and Structures*, 16(1), 47–56.

Campanile, L.F., 2006. 'Shape-adaptive wings - the unfulfilled dream of flight', in: R. Liebe (ed.), Flow Phenomena in Nature - A Challenge to Engineering Design, WIT Press, Southampton.

Campanile, L.F. and Anders, S., 2004. 'Aerodynamic and aeroelastic amplification in adaptive belt-rib airfoils', *Aerospace Science and Technology*, 9, 55–63.

Campanile, L.F. and Mierheim, O., 2005. 'Three-dimensional structural and aeroelastic analysis of morphing airfoils', Proc. 16th International Conference on Adaptive Structures and Technologies, Paris, France, 61–68.

Campanile, L.F. and Sachau, D., 2000. 'The belt-rib concept: a structronic approach to variable camber', *Journal of Intelligent Material Systems and Structures*, 11(3), 215–224.

Campanile, L.F., Sachau, D. and Seelecke, S., 1999. 'Ein adaptives Windkanalmodell für einen adaptiven Tragflügel', Proc. Deutscher Luft- und Raumfahrtkongress, Berlin, Germany, 19–28.

Campanile, L.F., Keimer, R. and Breitbach, E.J., 2004a. 'The "fish-mouth" actuator: design issues and test results', *Journal of Intelligent Material Systems and Structures*, 15(9–10), 711–719.

Campanile, L.F., Rose, M. and Breitbach, E.J., 2004b. 'Synthesis of flexible mechanisms for airfoil shape control: a modal procedure', Proc. 15th International Conference on Adaptive Structures and Technologies, Bar Harbour, ME, USA.

Canny, J., Risner, J. and Subramanian, V., 2002. 'Flexonics'. In: Boissonnat, J.D., Burdick, J., Goldberg, K., Hutchinson, S. (eds), Algorithmic Foundations of Robotics V, Springer Tracts in Advanced Robotics, 203–219.

Carpi, F., Migliore, A., Serra, E. *et al.*, 2005. 'Helical dielectric elastomer activators', *Smart Materials and Structures*, 14, 1210–1216.

Chattopadhyay, A., Seeley, C.E. and Jha, R., 1999. 'Aeroelastic tailoring using piezoelectric actuation and hybrid optimisation', *Smart Materials and Structures*, 8, 83–91.

Chopra, I., 2002. 'Review of state of art of smart structures and integrated systems', *AIAA Journal*, 40(11), 2145–2187.

Conway, H.D., 1956. 'The nonlinear bending of thin circular rods', *Journal of Applied Mechanics*, 7–10.

Crawley, E.F. and de Luis, J., 1987. 'Use of piezoelectric actuators as elements of intelligent structures', *AIAA Journal*, 25(10), 1373–1385.

De Bona, F. and Zelenika, S., 2002. 'Analytical and experimental characterisation of high-precision flexural pivots subjected to lateral loads', *Precision Engineering*, 26, 381–388.

Eastman, F.S., 1937. 'The design of flexure pivots', *Journal of the Aeronautical Sciences*, 5, 16–21.

Eppler, R., 1985. 'Vorrichtung zur Lagerung von Rudern und Wölbungsklappen von Flugzeugen und Wasserfahrzeugen', German Patent No. 2755442.

Euler, L., 1744. 'Methodus inveniendi lineas curvas maximi minimive proprietate gaudentes, sive solutio problematis isoperimetrici latissimo sensu accepti', Bousquet, Lausanne.

Frank, G.J., Joo, J.J., Sanders, B., Garner, D.M. and Murray, A.P., 2004. 'Mechanization of a high aspect ratio wing for aerodynamic control', Proc. 15th International Conference on Adaptive Structures and Technologies, Bar Harbour, ME, USA.

Frecker, M.I., 2003. 'Recent advances in optimisation of smart structures and actuators', *Journal of Intelligent Material Systems and Structures*, 14(4–5), 207–216.

Frecker, M.I., Ananthasuresh, G.K., Nishiwaki, S., Kikuchi, N. and Kota, S., 1997. 'Topological synthesis of compliant mechanisms using multi-criteria optimisation', *Journal of Mechanical Design*, 119, 238–245.

Freise, H., 1949. 'Federgelenke für Beanspruchungs- und Schwingungsmessgeräte', *Archiv für Technischer Messen*, V, 170.

Frisch-Fay, R., 1962. Flexible bars. Butterworth, Washington, DC.

Fu, Y., Du, H., Huang, W., Zhang, S. and Hu, M., 1994. 'TiNi-based thin films in MEMS applications: a review', *Sensors and Actuators A*, 112, 395–408.

Giurgiutiu, V., Pomirleanu, R. and Rogers, C.A., 2000. 'Energy-based comparison of solid-state actuators', Report No. USC-ME-LAMSS-2000-102, University of South Carolina, Columbia, SC.

Goldfarb, M. and Speich, J.E., 1999. 'A well-behaved revolute flexure joint for compliant mechanism design', *Journal of Mechanical Design*, 121(3), 424–429.

Gorski, W., 1976. 'A review of literature and a bibliography on finite elastic deflection of bars', *Civil Engineering Transactions*, 18(2), 74–85.

Gould, S.J., 1981. 'Kingdoms without wheels', *Natural History*, 90, 42–48.

Grant, C.H., 1935. 'Airfoil', US Patent No. 2,022,806.

Haberland, R., 1981. 'Flexure joint, particularly for connecting a gyroscope to its driving shaft', US Patent No. 4,261,211.

Haringx, J.A., 1949. 'The cross-spring pivot as a constructional element', *Applied Scientific research*, A1(5–6), 313–332.

Hart-Davis, A., 2006. 'Isambard Kingdom Brunel and the engineers of the nineteenth century', The Oxford Dictionary of National Biography, Oxford University Press.

Her, I. and Midha, A., 1987. 'A compliance number concept for compliant mechanisms and type synthesis', *Journal of Mechanisms, Transmissions, and Automation in Design*, 109(3), 348–355.

Hesselbach, J., Pittschellis, R., Hornbogen, E. and Mertmann, M., 1997. 'Shape memory alloys for use in miniature grippers', Proc. 2nd International Conference on Shape Memory and Superelastic Technologies, Pacific Grove, CA, USA, 251–256.

Hetrick, J.A. and Viota, S. 1999. 'An energy formulation for parametric size and shape optimization of compliant mechanisms', *Journal of Mechanical Design*, 121(2), 229–234.

Holle, A.A. and Judge, A.W., 1916. 'Improvements in and relating to Planes or the like for Aeroplanes', UK Patent Specification No. 15290.

Horie, M. and Ikegami, K., 1995. 'Development of pantograph mechanisms with super elastic hinges for micro-bonding by adhesives', Proc. 9th World Congress on the Theory of Machines and Mechanisms, Milan, Italy, 3096–3100.

Howell, L.L., 2001. Compliant Mechanisms, John Wiley & Sons, Inc. New York.

Howell, L.L. and Midha, A., 1994. 'A method for the design of compliant mechanisms with small-length flexural pivots', *Journal of Mechanical Design*, 116, 280–287.

Howell, L.L. and Midha, A., 1995. 'Determination of the degrees of freedom of compliant mechanisms using the pseudo-rigid-body model concept', Proc. 9th World Congress on the theory of machines and mechanisms, Milan, Italy, 1537–1541.

Howell, L.L. and Midha, A., 1996. A loop-closure theory for the analysis and synthesis of compliant mechanisms, *Journal of Mechanical Design*, 118, 121–125.

Huber, J.E., Fleck, N.A. and Ashby, M.F., 1997. 'The selection of mechanical actuators based on performance indices', *Proceedings of the Royal Society of London A*, 453, 2185–2205.

Ifju, P.G., Jenjins, D.A., Viieru, D. and Shyy, W., 2006. 'Flexible-wing-based micro air vehicles', in: R. Liebe (ed.), Flow Phenomena in Nature - A Challenge to Engineering Design, WIT Press, Southampton.

Jänker, P. and Martin, W., 1993. 'Performance and characteristics of actuator material'. Proc. 4th International Conference on Adaptive Structures, Cologne, Germany, 126–138.

Johnson, W.E., 1937. 'The load-deflection characteristics of initially curved flexural springs', *Journal of Applied Mechanics*, A119–A127.

Kiehl, M., 2006. 'Analyse, Entwurf und rechnerischer Nachweis einer formveränderlichen Tragfläche', DLR-IB 131-2006/28, 69.

Kota, S., 2001. 'Compliant systems using monolithic mechanisms', *Smart Materials Bulletin*, March, 7–10.

Kota, S., Hetrick, J. and Kikuchi, N., 1999a. 'Robustness of compliant mechanism topology optimization formulations', Proc. of SPIE Vol. 3667, 244–254.

Kota, S., Hetrick, J., Li, Z. and Saggere, L., 1999b. 'Tailoring unconventional actuators using compliant transmissions: design methods and applications', *IEEE/ASME Transactions in Mechatronics*, 4(4), 396–408.

Kota, S., Joo, J., Li, Z., Rodgers, S.M. and Sniegowski, J., 2001. 'Design of compliant mechanisms: application to MEMS', *Analog Integrated Circuits and Signal Processing*, 29, 7–15.

Kuribayashi, K., 1993. 'Criteria for the evaluation of new actuators as energy converters', *Advanced Robotics*, 7(4), 289–307.

La Barbera, M., 1983. 'Why the wheels won't go', *American Naturalist*, 121(3), 395–408.

Lazarus, K.B., Lundstrom, M.E., Moore, J.W., Crawley, E., Russo, F. and Yoshikawa, S., 2002. 'Packaged strain actuator', US Patent No. 6,404,107.

Lee, N. and Hanlon, J., 1977. 'Improvements relating to aircraft wings', UK Patent Specification No. 1 496 519.

Lilienthal, O., 1889. Der Vogelflug als Grundlage der Fliegekunst, R. Gaertners, Berlin, 185.

Lindhe Norberg, U.M., 2002. 'Structure, form, and function of flight in engineering and the living world', *Journal of Morphology*, 252, 52–81.

Lobontiu, N., 2003. Compliant Mechanisms: Design of flexure hinges. CRC Press, Boca Raton, FL.

Loewy, R.G., 1997. 'Recent developments in smart structures with aeronautical applications', *Smart Materials and Structures*, 6, R11–R42.

Loewy, R.G. and Tseng, S.P., 1993. 'Smart structures stabilized unstable control surfaces', AIAA Paper No. 93-1701-CP.

Lu, K.-J. and Kota, S., 2002. 'Compliant mechanism synthesis for shape-change applications: preliminary results', *Proc. SPIE*, 4693, 161–172.

Lu, K.-J. and Kota, S., 2003. 'Design of compliant mechanisms for morphing structural shapes', *Journal of Intelligent Material Systems and Structures*, 14(6), 379–391.

Lu, T.J., Hitchinson, J.W. and Ewans, A.G., 2001. 'Optimal design of a flexural actuator', *Journal of the Mechanics and Physics of Solids*, 49, 2071–2093.

Lyon, D.G., 1965. 'Variable shaped airfoil', US Patent No. 3,179,357.

Mankame, N,D. and Ananthasuresh, G.K., 2004. 'Topology synthesis of electrothermal compliant mechanisms using line elements', *Journal of Structural and Multidisciplinary Optimization*, 26, 209–218.

McKinney, M., 1977. 'Variable camber leading edge airfoil system', US Patent No. 4,040,579.

McKinney, M., 1982. 'Variable camber trailing edge for airfoil', US Patent No. 4,312,486.

Midha, A., Norton, T.W. and Howell, L.L., 1994. 'On the nomenclature, classification, and avbstractions of compliant mechanisms', *Journal of Mechanical Design*, 116, 270–279.

Nelli Silva, E.C., Nishiwaki, S. and Kikuchi, N., 2000. 'Topology optimisation design of flextensional actuators', *IEEE Transactions on Ultrasonics, Ferroelectrics, and Frequency Control*, 47(3), 657–671.

Padula, S.L. and Kincaid, R.K., 1999. 'Optimisation strategies for sensor and actuator placement', NASA/TM-1999-209126.

Paros, J.M. and Weisbord, L., 1965. 'How to design flexure hinges', *Machine Design*, 37, 151–156.

Pasini, D., 2004. 'Design optimisation of a solid-state hinge in a compliant mechanism', DLR-IB 131-2004/10, 55–62.

Parmele, W.R., 1931. 'Wing construction for aeroplanes', US Patent No. 1,803,915.

Pendleton, E.W., Bessette, D., Field, P.B., Miller, G.D. and Griffin, K.E., 2000. 'Active Aeroelastic Wing flight research program: technical program and model analytical development', *Journal of Aircraft*, 37(4), 554–561.

Perry, B., Cole, S.R. and Miller, G.D., 1995. 'Summary of an Active Flexible Wing program', *Journal of Aircraft*, 32(1), 10–15.

Piening, M. and Monner, H.P., 2000. 'Anströmprofil mit variabler Profiladaption', German Patent No. 19741490.

Pierce, D., 1978. 'Improvement in or related to aerofoils', UK Patent Specification No. 1 536 331.

Pini, N. and Ermanni, P. 2004. 'Patterned structures for AFCs applications made of thick layers of conductive polymers', Proc. 15th International Conference on Adaptive Structures and Technologies, Bar Harbour, ME, USA.

Ramrakhyani, D., Lesieutre, G.A., Frecker, M. and Bharti, S., 2005. 'Aircraft structural morphing using tendon-actuated compliant cellular trusses', *Journal of Aircraft*, 42(6), 1615–1621.

Rocheville, H.D., 1932. 'Airplane wing', US Patent No. 1,846,146.

Rolt, L.T.C., 1989, Isambard Kingdom Brunel, Penguin Books, Harmondsworth.

Rowarth, R., 1981. 'Variable camber wing', UK Patent Application No. 2 059 368.

Saggere, L. and Kota, S., 1999. 'Static shape control of smart structures using compliant mechanisms', *AIAA Journal*, 37(5), 572–578.

Schmidt, O., 1895. 'Steuerapparat für Flugmaschinen', German Patent No. 84532.

Sevak, N.M. and McLarnan, C.W., 1974. 'Optimal synthesis of flexible link mechanisms with large static deflections', *Journal of Engineering for Industry*, ASME-Paper No. 74-DET-83.

Shigley, J.E. and Uicker Jr, J.J., 1980. Theory of machines and mechanisms, McGraw-Hill, New York, 13.

Shoup, T.E. and McLarnan, C.W., 1971. 'A survey of flexible link mechanisms having lower pairs', *Journal of Mechanisms*, 6, 99–105.

Smith, S.T., 2000. Flexures: Elements of elastic mechanisms, CRC Press, Boca Raton, FL.

Spadoni, A. and Ruzzene, M., 2005. 'Static aeroelastic response of chiral-core airfoils', Proc. 16th International Conference on Adaptive Structures and Technologies, Paris, France, 190–198.

Statkus, F.D., 1982. 'Continuous skin, variable camber airfoil edge actuating mechanism', US Patent No. 4,351,502.

Staufenbiel, R., 2002. 'Laudatio zur Verleihung der Otto-Lilienthal-Medaille 2001 an Professor Dr.-Ing. Rolf Riccius', *DGLR-Mitteilungen*, 1, 1–4.

Strelec, J.K., Lagoudas, C.L., Khan, M.A. and Yen, J., 2003. 'Design and implementation of a shape memory alloy actuated reconfigurable airfoil', *Journal of Intelligent Material Systems and Structures*, 14(4–5), 257–273.

Tenzer, P.E. and Ben Mrad, R., 2004. 'On amplification in inchworm™ precision positioners', *Mechatronics*, 14(5), 515–531.

Trease, B.P., Moon, Y.-M. and Kota, S., 2005. 'Design of large-displacement compliant joints', *Journal of Mechanical Design*, 127(4), 788–798.

Troeger, H., 1962. 'Considerations in the applications of flexural pivots', *Automatic Control*, 17(4), 41–46.

Venanzi, S., Giesen, P. and Parenti-Castelli, V., 2005. 'A novel technique for position analysis of planar compliant mechanisms', *Mechanism and Machine Theory*, 40, 1224–1239.

Wei, Z.G., Sandström, R. and Miyazaki, S., 1998. 'Shape memory materials and hybrid composites for smart systems', *Journal of Materials Science*, 33, 3763–3783.

Wierach, P., 2003. 'Entwicklung multifunktionaler Werkstoffsysteme mit piezoelektrischen Folien im Leitprojekt Adaptronik', Proc. Adaptronic Congress 2003, Wolfsburg, Germany.

Wierach, P. and Schönecker, A., 2005. 'Bauweisen und Anwendungen von Piezocompositen in der Adaptronik', Proc. Adaptronic Congress 2005, Göttingen, Germany.

Wilkie, W.K., Bryant, R.G., High, J.W., Fox, R.L., Hellbaum, R.F., Jalink Jr, A., Little, B.D. and Mirick, P.H., 2000. 'Low-cost piezocomposite actuator for structural control applications', *Proc. SPIE*, 3991, 323–334.

Wilkie, W.K., Bryant, R.G., Fox, R.L., Hellbaum, R.F., High, J.W., Jalink Jr, A., Little, B.D. and Mirick, P.H., 2003. 'Method of fabricating a piezoelectric composite apparatus', US Patent No. 6,629,341.

Winzek, B., Schmitz, S., Rumpf, H., Sterzl, T., Hassdorf, R., Thienhaus, S., Feydt, J., Moske, M. and Quandt, E., 2004. 'Recent developments in shape memory thin film technology', *Materials Science and Engineering A*, 378, 40–46.

Wright, O., 1988a. How we invented the airplane. An illustrated history. Dover, New York, 14–15.

Wright, O., 1988b. How we invented the airplane. An illustrated history. Dover, New York, 12.

Wuest, W., 1950. 'Blattfedergelenke für Messgeräte', *Feinwerktechnik*, 54(7), 167–170.

Yin, L. and Ananthasuresh, G.K., 2003. 'A novel formulation for the design of distributed compliant mechanisms', *Mechanics Based Design of Structures and Machines*, 31(2), 151–179.

5

Adaptive Aeroelastic Structures

Jonathan E. Cooper

School of Mechanical, Aerospace and Civil Engineering, The University of Manchester, Manchester M60 1QD, UK

5.1 INTRODUCTION

Aeroelasticity [Bisplinghopf *et al.* 1996] is the science that considers the interaction of aerodynamic, elastic and inertial forces, as demonstrated by the classic Collar's Aeroelastic Triangle shown in Figure 5.1. Subsets of aeroelasticity combine pairs of these forces, such that structural dynamics results from the application of inertial forces on elastic bodies, and stability and control are due to the interaction of aerodynamic and inertial forces on a rigid body. Static aeroelastic phenomena occur when aerodynamic forces act upon a flexible body. For example, the aerodynamic forces acting upon a wing are dependent upon the wing shape; however, these forces also cause the wing to bend and twist, thus changing the aerodynamic forces, and so on. Dynamic aeroelastic elastic effects require the full interaction among all three of the forces. Such phenomena are not just restricted to aircraft, with aeroelasticity being an important consideration for the design of structures such as turbine blades, bridges, chimneys, Formula 1 racing cars, etc. Here we shall be interested only in aerospace structures.

In general, aeroelastic phenomena are undesirable. For instance, the effectiveness of a control surface reduces with increasing airspeed and increased

Adaptive Structures: Engineering Applications   Edited by D. Wagg, I. Bond, P. Weaver and M. Friswell
© 2007 John Wiley & Sons, Ltd

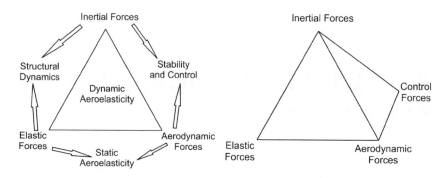

Figure 5.1   Collar's Aeroelastic Triangle and the Aeroservoelastic Pyramid

twist of the wing. In some cases complete structural failure can result from aeroelastic interactions such as flutter, a dynamic aeroelastic phenomenon, or divergence, which is a static aeroelastic effect. If an active control system is added to an aircraft, for instance to alleviate gust loads, then unfavourable interactions between the control system and the aeroelastic system can occur. This type of interaction is termed aeroservoelasticity (ASE), as shown in Figure 5.1.

It can be argued that aeroelasticity affected aircraft design even before the Wright brothers' first controlled and manned powered flight in 1903. Samuel Langley's 'Aerodrome' suffered structural failure and crashed into the Potomac River some months before the Wright brothers' first flight [Collar 1978], and photographs of this attempt show that divergence (a catastrophic static aeroelastic phenomenon) probably occurred. When a replica of the aircraft was built some years later but with stiffer wings, it flew. This is a good demonstration that structural stiffness is a very important parameter when aeroelastic design is considered.

Consequently, for nearly all of the 100 years since the beginning of powered flight, aeroelastic phenomena have had a significant influence upon aircraft structural design. In particular, as many prototype aircraft have been destroyed due to the occurrence of either flutter or divergence, it has been accepted that aircraft lifting surfaces have to be built to be stiff, and therefore heavy, enough so that neither phenomenon occurs within the desired flight envelope. Indeed, the most common 'fix' to deal with flutter problems that might arise within aircraft development programmes is to add extra mass to the structure. The consequent increase in structural weight has been called the 'aeroelastic penalty' [Kusmina *et al.* 2002].

There has been a recent drive towards improving aircraft performance. For the civil aircraft industry this has been focused not just upon reducing fuel costs, but also on environmental issues which are becoming of increasing importance. Aircraft manufacturers often refer to the 'green aircraft' of the future that will have much reduced emissions. The EU 20–20 Vision sets targets for the European aerospace industry of meeting a 50 % reduction in fuel usage and emissions by the year 2020. These targets cannot be met simply by the development of improved engines, lighter structures or the introduction of aerodynamic technologies such as flow control. An integrated approach will need to be taken in order to get much improved performance, and one of the key elements towards this will be improved use of aeroelastic technologies.

Consequently, there has been a recent change in focus as regards aeroelasticity, and researchers are starting to consider the possibility of using aeroelastic deflections to achieve gains in aerodynamic performance (whilst making sure that aeroelastic failure still cannot occur). Although the use of passive aeroelastic tailoring (employing unidirectional composite structures to influence aeroelastic behaviour) has been possible since the 1980s, any major improvements in flight performance due to aeroelastic effects will come from adaptive concepts. As with many technical areas, such approaches have been considered before: Lilienthal employed active camber control on some of his gliders and the Wright brothers used wing warping as the mechanism for roll control; however, it is only recently that adaptive aeroelastic concepts have been reconsidered [Pendleton *et al.* 2000, Pendleton 2001, Perry *et al.* 1995, Flick and Love 1999, Schweiger and Krammer 1999].

Adaptive concepts need to use aeroelastic effects to get the full benefit of the deflections throughout the entire flight envelope. Figure 5.2 shows how aeroelastic phenomena have limited possible aerodynamic performance since

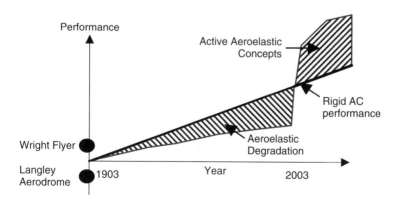

Figure 5.2 Effect of aeroelasticity on aircraft performance (courtesy of Hans Schweiger)

the start of manned powered flight, and how the use of adaptive aeroelastic concepts could make a dramatic step change.

Such concepts are likely to be used for a range of different applications, including:

- Drag reduction – by changing the wing or camber shape, or winglet orientation, throughout the flight it will be possible to maintain the optimal lift/drag ratio at all points in the flight envelope and thus optimise the aircraft range.
- Load control – through the use of active changes in wing shape it will be possible to reduce the loads that are induced from gusts or manoeuvres without the need for control surfaces. Reduced loads imply that less structure is required, leading to better aircraft performance.
- Roll control – changes in the wing shape could be used for roll manoeuvres without the need (or reduced need) for control surfaces. Control-free designs are very important in terms of stealth, but reducing the size of control surfaces can have an important benefit in terms of performance.

Aircraft are traditionally designed to have optimal aerodynamic characteristics (maximum lift/drag ratio) for a single point in the flight envelope with no possibility of adapting the wing shape except through the use of high-lift devices for take-off and landing. However, the fuel loading and distribution change continuously throughout the flight and aircraft often have to fly at non-optimal flight conditions due to air traffic control restrictions. It is likely that environmental considerations will influence flight altitude and speeds to a greater degree in the future. This sub-optimal performance is much more significant for commercial aircraft as they are more flexible than military aircraft and where fuel efficiency is of far greater importance.

In recent years there has been a growing interest in the development of aircraft structures that allow aeroelastic deflections to be used in a beneficial manner and to enhance aerodynamic performance [Pendleton 2001]. For instance, the Active Flexible Wing [Perry et al. 1995], Active Aeroelastic Wing [Pendleton et al. 2000, Heeg et al. 2005] Programs investigated the use of using leading and trailing edge control surfaces to control the wing shape. The Morphing Program [Wlezien et al. 1998] developed a number of active aeroelastic concepts based upon smart materials and structures. In Europe, the 3AS (Active Aeroelastic Aircraft Structures) research project [Schweiger and Suleman 2003] also developed and demonstrated various active aeroelastic concepts, primarily in the areas of adaptive attachments, three surface aeroelastic aircraft and novel aeroelastic leading edge and wing tip devices. The Variable Stiffness Spar (VSS) approach [Florance *et al.* 2004, Chen *et al.* 2000] demonstrated the use of rotating spars for roll control.

The concept of adapting the shape of an aircraft in flight has been given the generic title of morphing; however, the activity can be divided clearly into two different categories [Bae *et al.* 2005]:

1. *Planform morphing* – where the aircraft planform is changed. Recent work [Bae *et al.* 2005] has investigated the use of telescopic, deployable and variable sweepback wings using a range of mechanical and pneumatic devices. One of the main drivers behind the use of this technology is the capability of changing mission mid-flight, e.g. the development of UCAVs (Uninhabited Combat Air Vehicles) that are able to loiter with high-aspect-ratio wings, but can then change their aspect ratio as they change to an attack role. Other examples have included the use of telescopic wings for roll control. Of course, there are many examples of military aircraft that have flown for many years with variable sweep wings.

2. *Performance morphing* – where the lifting surface planform remains the same but the aerodynamic shape (and hence performance) can be changed either through adjustments in the twist and bending behaviour along the wingspan, or through changes in the camber and/or the leading and trailing edge shapes. Such a capability could be used to maintain the best possible lift/drag ratio throughout the flight. However, there is also the possibility of implementing roll control (analogous to the Wright brothers' wing warping) which has gained some significance recently with the interest in control-surface-free UAVs (Unmanned Air Vehicles) in order to improve observability characteristics. Other work has examined the use of smart materials (e.g. piezo and shape memory alloys) to achieve this goal [Wlezien *et al.* 1998, Cesnik and Brown 2003], but there is still a lot of work and material development required in order to develop the considerable forces required to twist and bend (and maintain that deflection) a wing during flight.

In this chapter, two adaptive aeroelastic structures concepts are discussed that were developed as part of the 3AS research project [Schweiger and Suleman 2003]. The adaptive internal structures concept for control of the wing static aeroelastic shape in-flight is introduced and illustrated with two different approaches: changing the chord-wise position of the spars, and rotating the spars. The use of adaptive stiffness attachment devices to improve the effectiveness of all-moving vertical tails is also investigated. In both cases, the underlying philosophy behind the concepts, and the design and test of wind tunnel model prototypes, are described. Finally, the feasibility of applying the adaptive internal structures concept to full-size aircraft is discussed.

5.2   ADAPTIVE INTERNAL STRUCTURES

Adaptive aeroelastic concepts aim to control the shape of aircraft lifting surfaces in order to determine the lift and drag of the aerodynamic forces. The key idea exploited in the adaptive internal structures approach is to use the aerodynamic forces acting upon the wing to provide the forces and moments to bend and twist the wing, rather than trying to apply the forces via some form of actuator. Consider the schematic of the wing shown in Figure 5.3, with the lift acting at the aerodynamic centre on the quarter chord. By changing the position of the shear centre of the wing, the bending moment, and hence the amount of twist, will also change. It is envisaged that a far smaller amount of energy is required to adjust the structure compared with that required to twist the wing and to maintain the shape.

Such an approach is very attractive for adaptive aeroelastic concepts, where the lifting surface deflections could be adjusted gradually throughout the flight to maintain an optimal aerodynamic performance (i.e. maximise the lift/drag ratio). The concept could also be applied for roll control and gust or manoeuvre load alleviation; however, there would be a far greater requirement in these cases for the structural stiffness to change rapidly.

It is envisaged that the adaptive internal structures concept is not suitable for high-frequency applications (e.g. flutter suppression). It should also be noted that the approach is best applied to the tip end of the lifting surface, where the structural stiffness is far less than towards the wing root, and also the influence upon the aerodynamic forces is greatest. Also, although the examples shown in the rest of this section deal with full span adaptive spars, it should be remembered that only some elements of the wing internal structure would need to be adaptive. There are a number of different

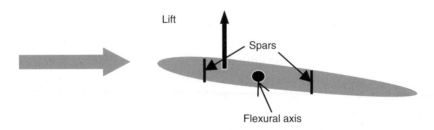

Figure 5.3   Components affecting wing shape

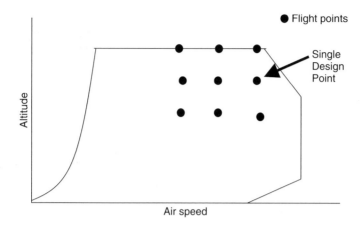

Figure 5.4 Typical flight envelope for civil aircraft

ways that the wing stiffness and shear centre position can be changed, but an important consideration that must be accounted for is the practicalities of any design: it must be ensured that the load can be transferred throughout the wing structure during all possible configurations and flight conditions.

Considering the typical flight envelope shown in Figure 5.4, aircraft are traditionally designed so that they have an optimal lift/drag ratio at a single point in the flight envelope (say mid-Atlantic for an aircraft designed to have a range from Europe to the USA). With the changing fuel loading throughout the flight, the wing shape will change. Air traffic control restrictions can also force the aircraft to fly at sub-optimal altitudes or speeds. Consequently, adaptive aeroelastic wings could be used to ensure that minimum drag is achieved throughout the flight whilst maintaining the requisite amount of lift.

The results of the initial studies shown here were part of an investigation to demonstrate that concepts based upon translation and rotation of the spars were feasible.

5.2.1 Moving Spars

Consider the schematic of the simple rectangular wing box shown in Figure 5.5 where the two thickest spars are able to move in a translational (chord-wise manner). The torsional stiffness is greatest when they are furthest apart and least when they are closest together. Note that the bending stiffness does not vary with position of the spars.

Figure 5.5   Moving spars concept: left, high torsional stiffness; right, low torsional stiffness

Now consider the aeroelastic analysis of a simple high aspect ratio rect-angular wing in order to demonstrate how the static and dynamic aeroelastic behaviour can be controlled through translational movement of the spars. As the purpose is to demonstrate the concept, only a rudimentary aeroelastic analysis has been employed. Modified strip theory aerodynamics was imple-mented along with thin walled structural theory and a Rayleigh-Ritz assumed shapes approach in order to study the static and aeroelastic behaviour of the wing, considering only the wing box itself. Of interest is the variation of the torsion constant, the position of the shear centre, the static twist, and also the effect upon the flutter and divergence speeds in relation to the position or orientation of the spars. All of these parameters can be calculated explicitly using this analytical approach, which is not the case if a finite element model were used.

A schematic of the proposed concept is shown in Figure 5.6. The simple rectangular wing box is made up of three identical uniform spars. The outer two spars remain in the same position, whereas the middle spar can be moved anywhere within the wing box. By moving the chord-wise position of the central spar, it is possible to change the torsional stiffness and also the position of the local shear centre, but the bending stiffness remains constant.

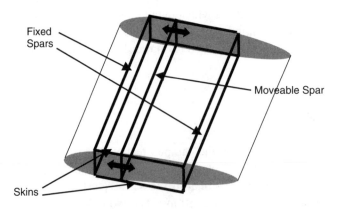

Figure 5.6   Three-spar rectangular wing with moveable middle spar

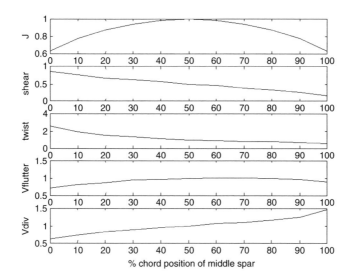

Figure 5.7   Effect of moving the middle spar

Figure 5.7 shows the changes in the torsion constant, shear centre position, wing tip twist, flutter and divergence speeds with respect to the position of the middle spar for the moving spar concept. All the results have been normalised with respect to the values found with the spar in the central position. These results demonstrate how the aeroelastic twist can be controlled simply by moving the spar. However, the twist is not simply a linear relation of the spar position because, although the torsion constant and shear centre position have the expected symmetric behaviour with the maximum value occurring when the spar is placed at the central position, the aerodynamic moment is related to the distance of the shear centre from the aerodynamic centre at the quarter chord.

The aeroelastic characteristics are dependent upon not just the torsional and bending (constant in this case) stiffness, but also the distance between the flexural axis and the aerodynamic centre. Although the bending stiffness does not change, there is still a coupling between the bending and torsion behaviour in the aeroelastic model. Inspection of the bottom three plots in Figure 5.7 demonstrates the relative complexity of this behaviour even for such a simple case. As the spar is moved from forwards to aft, the twist of the wing reduces. The flutter speed increases as the spar is moved aft until it reaches a maximum at a position of around 70 % and then it reduces. However, the divergence speed can be seen to rise steadily as the middle spar is moved from one side to the other.

A number of prototype demonstrators were designed and constructed via the use of several finite element models. The moving spar model differs slightly from the model described in the above numerical example in that there were two fixed outer spars and two inner spars whose chord-wise position could be varied. All of the wind tunnel models were rectangular, untapered wings with semi-span 0.775 m and chord 0.25 m. Being an aeroelastic study, the models were designed to be reasonably flexible and to fit in the test section of one of the low-speed wind tunnels at The University of Manchester.

Two versions of the moving spar concept were assessed, one where the position of the moving spars was controlled using a worm-drive driven by an electric motor, and in the other, a pneumatic ram. The main structure (see Figure 5.8) was constructed of aluminium and comprised five ribs, two fixed spars positioned horizontally near the leading and trailing edges, and two moveable spars positioned vertically. The moveable spars could translate in the chord-wise direction along the tracks in the ribs. For the purpose of the wind tunnel testing, specially formed balsa blocks were used to form the leading and trailing edges, and a thin polyethylene skin was added to provide the aerodynamic surface.

Static tests were performed by placing weights on the structure in order to demonstrate that it was feasible to move the spars whilst under load. It was found that whereas this was possible using the worm-drive, the pneumatic

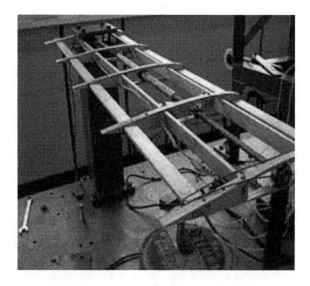

Figure 5.8 Prototype of the moving spar wind tunnel model

Figure 5.9 Maximum and minimum deflections achieved using the moving spar concept

ram did not have enough power and was discarded from the investigation. Figure 5.9 shows the deflections that were achieved, which differed in magnitude by around 30 mm and 2° twist at 25 m/s. It was possible to go from maximum to minimum stiffness in about 12 seconds.

5.2.2 Rotating Spars

In the second approach that was considered, the spars were able to rotate, thereby changing their bending stiffness in both vertical and horizontal directions, as well as the torsional stiffness and shear centre position of the wing section. Figure 5.10 shows the maximum and minimum stiffness cases.

The rotating spar concept was modelled in a similar way to the moving spar case described above with a two-spar wing box model but now with both

Figure 5.10 Rotating spar concept – maximum and minimum bending stiffness

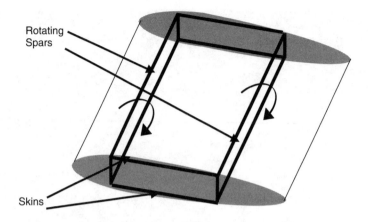

Figure 5.11 Two-spar rotating spar model

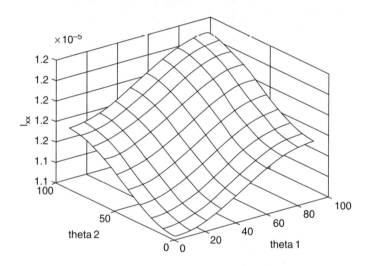

Figure 5.12 I_{xx} variation with leading and trailing edge rotation angle

the leading and trailing edge spars being able to rotate between 0° and 90° as shown in Figure 5.11. Figure 5.12 shows how the second moment of area varies with the orientation of both spars. Similar plots can be drawn for the torsion constant, shear centre position, wing tip twist, flutter and divergence speeds. The second moment of area and torsion constant values are directly related to the orientation of the spars and are greater the closer that they are to 90°. The position of the shear centre depends upon the relative rotation of the spars as it moves towards the stiffest one. As with the first concept, the behaviour of the tip twist, flutter and divergence speeds is not simple to

Figure 5.13 Prototype of the rotating spar wind tunnel model

predict, for, although a reduced torsional stiffness leads to larger deflections and reduced flutter speeds, this is complicated by the position of the shear centre relative to the aerodynamic centre.

As with the moving spar case, a prototype wind tunnel model of this concept was designed, manufactured and tested and is shown in Figure 5.13. The rotation of the spars was controlled via four motors positioned at either end of the spars and then attached to the ribs. It took about 0.5 seconds to rotate the spars from minimum to maximum stiffness. Note that for both of the concepts shown, and unlike a traditional aircraft wing design, the aerodynamic loads are transmitted from the skins to the ribs and then to the moving or rotating spars. Care has to be taken to ensure that these load paths are followed and that the moving/rotating spars do carry the load in all possible positions/orientations.

Experimentally measured bending and twist deflections are shown in Figures 5.14 and 5.15. The wing was much stiffer in torsion than in bending and hence the amount of twist is relatively small. The maximum twist is found when the leading edge spar has minimum stiffness and the trailing edge spar has maximum stiffness. The opposite orientation gives the minimum twist.

Further numerical studies have investigated the amount of lift and drag that result from a finite element model such as that shown in Figure 5.16 with all segments of the spar being able to be orientated between 0° and 90° (unlike the wind tunnel model where the orientation is constant along the entire span). A typical lift vs. drag trend is shown in Figure 5.17 for all combinations of spar segment rotation and it can be seen that different

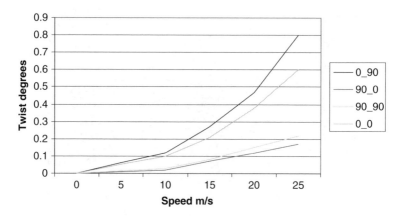

Figure 5.14 Wind tunnel twist for rotating spar model

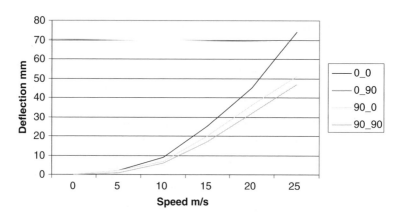

Figure 5.15 Wind tunnel bending deflection for rotating spar model

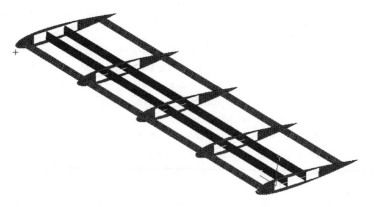

Figure 5.16 Typical finite element model of wind tunnel prototype

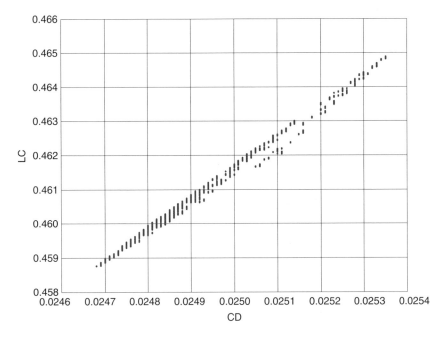

Figure 5.17   Lift coefficient vs. drag coefficient for rotating spar model

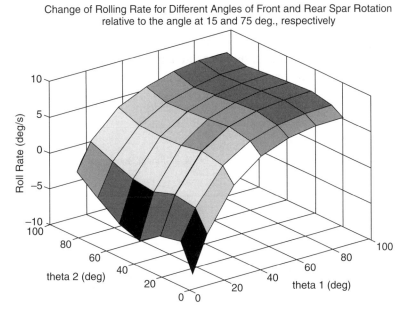

Figure 5.18   Rolling rate due to various orientations of leading and trailing edge spars

drag values can be found for the same lift. It is therefore possible to adjust the spar orientation in order to minimise the drag at all points in the flight envelope.

Finally, the rotating spar concept can be used to control the rolling rate. By changing the lift on each wing, but ensuring that the total lift remains the same, it is possible to roll the aircraft. Numerical simulations on a UAV-type aircraft produce the typical results shown in Figure 5.18 where the non-linear nature of the resulting roll rate in relation to the spar rotation can be clearly seen.

5.3   ADAPTIVE STIFFNESS ATTACHMENTS

The vertical lifting surfaces of high-speed aircraft suffer from reduced stability and control effectiveness at high dynamic pressures due to aeroelastic effects. Conventional designs ensure an adequate tail performance by having a large high-aspect-ratio fin, but these are prone to high loads. Consequently, such structures need to be very stiff, resulting in heavy structures. The consequent size and structural requirements lead to further weight and drag penalties.

Part of the 3AS project aimed at developing active aeroelastic all-moving vertical tails (AAAMVTs) with the target of producing improved fin designs with decreased tail size and structural weight whilst meeting, or exceeding, all tail performance goals. A key characteristic, as shown in Figure 5.19, of all-moving vertical tails is that they are attached by a single attachment. Such a concept is not new, and there are a number of previous aircraft that

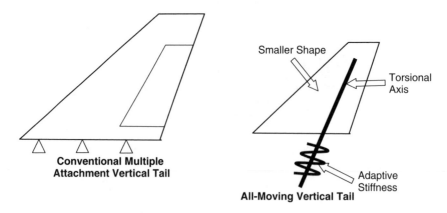

Figure 5.19   Conventional and all-moving fins

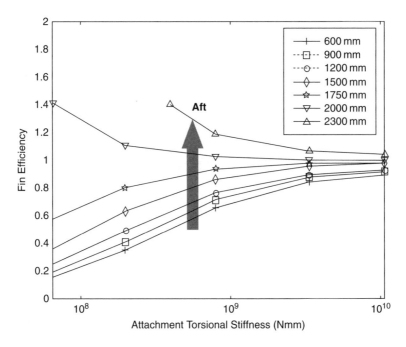

Figure 5.20   Aeroelastic effectiveness of typical all-moving vertical tail

have made use of them, e.g. VSR71 Blackbird, F117 Stealth Fighter, and also TSR2.

Initial studies on AAAMVTs have shown that the use of a single attachment enables a greater influence on the aeroelastic behaviour to be made than for a fin with multiple attachments. In particular, the attachment can be moved far backwards because there is no rudder and the fin can be utilized for the low-speed regime (engine out or side wind requirement) where there is no aeroelastic effect because the whole surface can be rotated. Figure 5.20 show typical values of the aeroelastic effectiveness for different attachment stiffness and different attachment positions for a range of Mach numbers. Once the attachment is moved far enough aft, it is possible to get an aeroelastic effectiveness greater than unity. This effect can be achieved for all Mach numbers if the stiffness is reduced sufficiently. The improvement over the rigid effectiveness is due to the positioning of the flexural axis relative to the aerodynamic centre.

However, it must not be forgotten that as the torsional stiffness is reduced, the flutter and divergence speeds will also reduce. Figure 5.21 shows an example of how the flutter speed reduces dramatically with diminishing stiffness and also as the attachment point is moved backwards. Consequently,

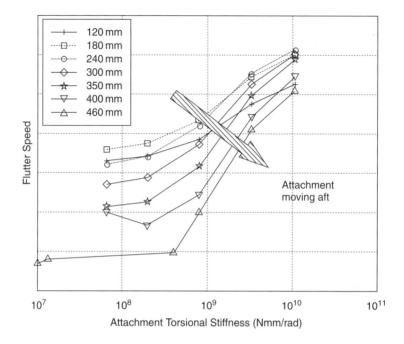

Figure 5.21   Flutter speed of typical all-moving vertical tail

there has to be a trade-off between the gains in the aeroelastic efficiency and aeroelastic stability considerations in any design. Any variable stiffness attachment is likely to be required to cope with the different requirements at low- and high-speed conditions; the only way to achieve this is through the use of an attachment with an adaptive torsional stiffness.

For high speeds, the vertical tail is designed to provide a certain minimum value of the directional static stability derivative. At low speeds, the rudder power must be adequate to cover the sideslip case for a cross-wind landing and also the one-engine-out case. This low-speed requirement may reduce the possibility to cut the fin span and area commensurate with positive high-speed aeroelastics. These requirements are more or less the same for both commercial and fighter aircraft.

As part of the 3AS project, an adaptive stiffness attachment employing pneumatic cylinders was developed for use with the all-moving vertical tail component, shown in Figure 5.22, that was designed to be demonstrated on the EuRAM wind tunnel model (7 m wing span) shown in Figure 5.23. Both the EuRAM model and the all-moving vertical tail were designed and manufactured by TsAGI. The goal was to produce an attachment that could

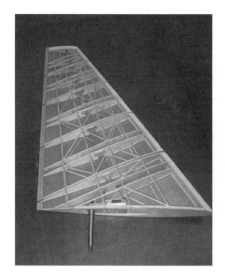

Figure 5.22   All-moving vertical tail

Figure 5.23   EuRAM wind tunnel model demonstrator

provide a maximum effectiveness of 1.5 whilst also being able to control the stiffness (and hence the effectiveness).

The work at Manchester investigated a number of different concepts, and the one described here made use of pneumatic cylinders whose effective stiffness can be altered via changes in the compressed air supply. Figure 5.24 shows the configuration that was used in this work. A pair of cylinders, attached in series to the air supply so that they both experience the same

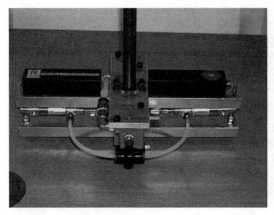

Figure 5.24   Pneumatic adaptive stiffness attachment and installation on EuRAM model

pressure, are used to resist the torsional motion. As the pressure is increased, the greater the resulting torsional stiffness becomes. The device that was manufactured was designed to operate up to a pressure of 10 bar, but due to limitations in the available air supply, the maximum possible air pressure was 6 bar.

It should be noted that the design of the pneumatic attachment, and the other attachments developed as part of the 3AS project, were constrained by the available space in the EuRAM model. If there had been a greater allowable space to position the attachment, then a different configuration would have been used. Of particular concern with the set-up used here was the fear that free play might become a problem, especially around the 0° position.

Initial testing of the device focused upon bench-tests of the adaptive device, in particular the torsional stiffnesses that could be achieved and also the practicalities of changing and controlling the stiffness. Different bending moments were applied and the rotation measured. These results allowed the torsional stiffness to be calculated for various pressures, thus enabling a calibration factor to be developed relating air pressure to stiffness. Figure 5.25 shows the resulting twist angles and demonstrates the linearity of the device.

Computational predictions of the EuRAM wind tunnel model showed that in order to achieve the required effectiveness of 1.5, whilst ensuring that flutter and divergence would not occur, a rotational stiffness of 70 N m/rad needed to be achieved. Inspection of Figure 5.25 shows that this requirement was met.

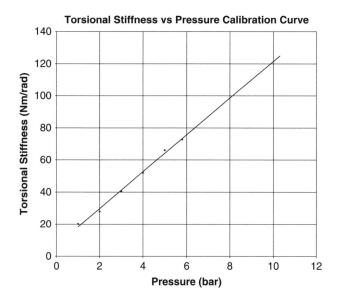

Figure 5.25 Torsional deflections for applied bending moments

It was of great interest to investigate the practicality of changing and controlling the air pressure and hence the torsional stiffness. This turned out to be relatively straightforward, with changes in the stiffness being almost instantaneous. A controller device was designed and manufactured that enabled the air pressure to be changed to meet the required torsional stiffness and its use was very successful and easy. It was possible to go from zero stiffness to maximum stiffness (or maximum to zero) in less than half a second. When implemented in the wind tunnel tests there was a certain amount of overswing due to the inertial effects of the rotating vertical tail component, but this was damped out in a few cycles.

Modal testing of the vertical tail with adaptive stiffness device was performed using a hammer test. As was to be expected, the flexible modes of the AAAMVT remained more or less the same for different attachment stiffnesses, but the underlying low-frequency rigid-body torsion mode was directly related to the stiffness of the attachment. These results were used to validate a finite element model of the fin with adaptive attachment.

A series of static and dynamic tests were then performed in the environmental wind tunnel at the Goldstein Laboratories in Manchester. Two laser displacement devices were used to measure the response for all of the tests. A range of different torsional stiffnesses were implemented for several different attachment positions.

Static testing consisted of setting the AAAMVT at several initial angles of attack and measuring the static twist angle for differing speeds. The effectiveness was defined as the ratio of the resulting total twist angle (including the initial angle) to the initial angle. This process was repeated for several torsional stiffnesses at the different attachment positions. Figure 5.26 shows some of the effectiveness measurements for the fin at different wind tunnel speeds with a rearward (in this case 50 % mean aerodynamic chord) attachment. As predicted from the aeroelastic analysis, the effectiveness is greater than unity and increases with speed and decreasing torsional stiffness.

Dynamic tests were then performed to determine how the vibration characteristics for different parameter settings (torsional stiffness, attachment position) varied with tunnel speed. A chord was used to provide a step input for the system and the resulting vibration decay was measured. This time history was then curve-fitted using a time domain system identification algorithm in order to determine the natural frequencies and damping ratios. These results compared reasonably well with aeroelastic predictions.

A final test was used to investigate the possible free-play characteristics of the adaptive stiffness attachment. At a particular stiffness setting, the rotating turntable of the environmental tunnel was moved from +15° through to −15°. Such a procedure meant that the AAAMVT changed the side to which it was twisted. By examining the resulting angle of twist, it is possible

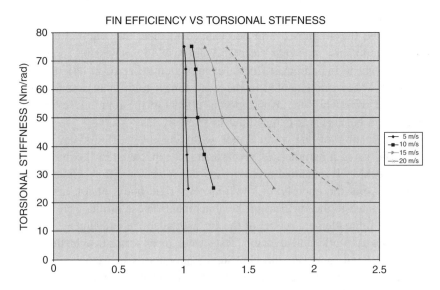

Figure 5.26   Fin effectiveness measurements for different torsional stiffness attachments

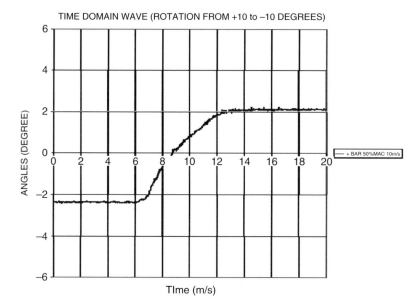

Figure 5.27   Free-play investigation results

to see whether any free play is present as there would be a sudden change in angle of twist should that be the case. Figure 5.27 shows that the initial aeroelastic twist was just greater than $-2°$ and this changes to just over $2°$ once the turntable has rotated. The twist angle in the region where the turntable is turning is effectively linear, there are no sudden step changes in the deflection, and consequently any free play in the system was negligible.

Following these component tests, the vertical tail and pneumatic device were sent to Moscow and successfully tested by TsAGI on the EuRAM model.

5.4 CONCLUSIONS

It has been shown that through the use of adaptive aeroelastic structures it is possible to control and adjust the aerodynamic performance of aircraft lifting surfaces influencing the drag, roll control and aeroelastic effectiveness. Two different approaches, namely adaptive internal structures and adaptive stiffness attachments, have demonstrated successfully, using numerical aeroelastic analysis and also prototype wind tunnel models, that it is possible to control the aeroelastic behaviour of the lifting surfaces. The experimental investigations showed that the practical implementation is not straightforward as it must be ensured that the aerodynamic loads are carried through the adaptive structures, and a great deal of thought needs to be given to achieve effective designs.

5.5   THE WAY FORWARD

These initial studies have shown that it is possible to control the aerodynamic performance of commercial aircraft through the use of adaptive aeroelastic structures technology. In the future it is conceivable that aircraft could fly with adaptive aeroelastic and sensing systems on-board that enable the wing (including winglets and tail surfaces) shape and deflection to be adjusted continually in order to:

- ensure optimum lift/drag ratio throughout the entire flight envelope;
- facilitate roll control with reduced size control surfaces;
- reduce the need for large high-lift surfaces for take-off and landing;
- reduce the noise on take-off and landing;
- enable improved load control and reduction due to gusts and manoeuvres.

However, a great deal of further work is required before such concepts could be used by the aerospace industry. Research is currently underway to investigate the location on the structure of the most efficient place to adjust the structural stiffness, exactly how much change in stiffness is required and also the best way of achieving these changes in stiffness. It is not envisaged that the entire wing structure would adapt; for instance, the main spars of the A380 are larger than a human being and there is no sense in attempting to adapt such massive structures. There is much greater sense in applying the adaptive structures technology to the wing tips where the wing structure is much smaller due to the reduced loads and also where changes in the chord-wise wing shape have a greater effect on the aerodynamics than further in-board parts of the wing.

The power and weight requirements as well as full-scale implementation issues of any proposed adaptive aeroelastic structure are very important because there is no point in introducing such technologies if the gains in the aerodynamic performance are outweighed by such penalties. The moving and rotating spar concepts described above depend upon a wing structure that does not have these adaptive spars attached to the outer skins and this will require either more ribs or thicker skins. Fuel is usually stored in the wing and the implementation of adaptive structures will cause difficulties in the design. Consequently, trade-off studies need to be performed in order to establish the pros and cons as well as determining the best way to incorporate the concepts into aircraft structural designs.

It is likely that the first implementations of adaptive aeroelastic structures will be in UAVs. This is due to the economies of scale and in particular the reduced forces that are required. There is a current research interest, driven by stealth requirements, in developing flapless UAVs where adaptive aeroelastic structures could make a big contribution, possibly in combination with flow control technologies. There is also much interest in HALE (High-Altitude Long-Endurance) or Sensorcraft UAVs where the extreme flexibility of very high-aspect-ratio structures could benefit from the use of adaptive stiffness technologies.

All of the concepts considered in this chapter have used innovative implementations of traditional technologies. There is much scope for the use of smart materials, such as shape memory alloys, and structures in the implementation of adaptive aeroelastic structures; however, the current performance of such materials is not yet at a stage where full scale implementation could be contemplated.

Finally, despite the encouraging prospects of adaptive aeroelastic structures, it is unlikely that the far-reaching goals of improved aerodynamic performance and load reduction are achievable solely with the use of adaptive aeroelastic structures, and the implementation in combination with other technologies such as flow control needs to be investigated.

ACKNOWLEDGEMENTS

The author is grateful for the contribution of Mike Amprikidis, Gareth Vio, Vijaya Hodigere, Otto Sensburg, Hans Schweiger, Svetlana Kuzmina, Mikhail Zichenkov and all other colleagues from the 3AS project to this work.

Part of the work described was undertaken during the academic year 2005–2006 when the author was a Royal Academy of Engineering/Leverhulme Trust Senior Research Fellow.

Part of this work was funded by the European Union as part of the Active Aeroelastic Aircraft Structures Research Programme. Active Aeroelastic Aircraft Structures (3AS) is a research project partially funded by the European Union under the New Perspectives in Aeronautics Key Action of the 'Competitive and Sustainable Growth' Framework Programme 5. Partners in the project are: EADS-Deutschland, Alenia, EADS-CASA, GAMESA, Saab, CIRA, DLR, INTA, VZLU, KTH, IST, The University of Manchester, Politecnico di Milano, Technion and TsAGI.

REFERENCES

Bae, Jae-Sung, Seigler, T. Michael & Inman, D.J., 2005 'Aerodynamic and Static Aeroelastic Characteristics of a Variable-Span Morphing Wing', J. Aircr. v42 n2 pp 528–534.

Bisplinghof, R., Ashley, H. & Halfman, R., 1996 'Aeroelasticity', Dover (original 1955).

Cesnik, C.E.S. & Brown, E.L., 2003 'Active Warping Control of a Joined-Wing Airplane Configuration', 44th AIAA Structures, Structural Dynamics and Materials Conference, AIAA 2003–1715.

Chen, P.C., Sarhaddi, D., Jha, R., Lui, D.D., Griffin, K. & Yurkovich, R., 2000 'Variable Stiffness Spar Approach for Aircraft Manoeuvre Enhancement Using ASTROS', J. Aircr. v37 n5 pp 865–871.

Collar, A.R., 1978 'The First Fifty Years of Aeroelasticity', Aerospace Feb, pp 12–20.

Flick, P. & Love, M., 1999 'The Impact of Active Aeroelastic Wing Technology on Conceptual Aircraft Design', AVT Panel Meeting, Ottawa.

Florance, J.R., Heeg, J., Spain, C.V., Ivanco, T.G., Wieseman, C.D. & Lively, P.S., 2004 'Variable Stiffness Spar Wind-Tunnel Model Development and Testing', 45th AIAA Structures, Structural Dynamics and Materials Conference, AIAA 2004–1588.

Heeg, J., Spain, C.V., Florance, J.R., Wisemann, C.D., Ivanco, T.G., DeMoss, J.A. & Silva, W.A., 2005 'Experimental Results from the Active Aeroelastic Wing Wind Tunnel Test Program', 46th AIAA Structures, Structural Dynamics and Materials Conference, AIAA 2005–2234.

Kusmina, S. et al., 2002 'Review and Outlook for Active and Passive Aeroelastic Design Concepts for Future Aircraft', International Conference on Aeronautical Structures.

Pendleton, E., 2001 'Back to the Future - How Active Aeroelastic Wings are a Return to Aviation's Beginnings and a Small Step to Future Bird-like Wings', RTO-SMP Panel Meeting.

Pendleton, E., Bessette, D., Field, P., Miller, G. & Griffen, K., 2000 'Active Aeroelastic Wing Flight Research Program and Model Analytical Development', J. Aircr. v37 n4 pp 554–561.

Perry, R., Cole, S.R. & Miller, G.D., 1995 'Summary of an Active Flexible Wing Program', J. Aircr. v32 n1 pp 10–15.

Schweiger, J. & Suleman, A., 2003 'The European Research Project – Active Aeroelastic Structures', CEAS International Forum on Aeroelasticity and Structural Dynamics.

Schwiger, D. & Krammer, J., 1999 'Active Aeroelastic Aircraft and its Impact on Structure and Flight Control System Design', AVT Panel Meeting, Ottawa.

Wlezien, R.W., Horner, G.C., McGowan, A.R., Padula, S.L., Scott, M.A., Silcox, R.J. & Simpson, J.O., 1998 'The Aircraft Morphing Program', SPIE Smart Structures and Materials Meeting, pp 176–187.

6

Adaptive Aerospace Structures with Smart Technologies – A Retrospective and Future View

Christian Boller

The University of Sheffield, Department of Mechanical Engineering, Sheffield S1 3JD, UK

6.1 INTRODUCTION

When smart structures technologies emerged around two decades ago expectations were high as usual. Many of the ideas were generated from space vehicle applications, which gave rise to a lot of spin-offs and which had to be specified. Large visions were outlined starting mainly from different NASA laboratories, which then quickly spread to Japan and Europe, and with the latter to the UK and Germany. While the UK looked at a broader view with a range from sports articles to aerospace, initially reflected by B. Culshaw through the foundation of the *Smart Structures Research Institute* at Strathclyde University in 1991 and followed by the *EPSRC Smart Technologies Strategic Workshop* in Loch Lomond in 1997 organised by G.R. Tomlinson

Adaptive Structures: Engineering Applications   Edited by D. Wagg, I. Bond, P. Weaver and M. Friswell
© 2007 John Wiley & Sons, Ltd

from the University of Sheffield, the German side worked the other way around, starting from the narrower field of aerospace and then gradually expanding this into the automotive, machining and medical sectors. A key strategic paper referenced many times in that regard was the one published by E. Breitbach [1] in 1991. The views communicated there with respect to aerospace included:

- *Fixed wing aircraft:*

  - Attenuation of dynamic loads by means of an active wing–fuselage interface (based on piezoceramics).
  - Flutter suppression and vibration reduction by means of adaptive stiffness tuning and adaptive control of the wing camber (in these cases the applications of shape memory alloys, SMAs, were seen to be efficient).
  - Active internal noise cancellation.
  - Adaptive stiffness control.
  - Adaptive wing–engine pylon.

- *Rotary wing aircraft:*

  - Adaptive blade roots with embedded multifunctional actuators.
  - Adaptive control rods designed as active systems with built-in smart actuators (higher harmonic control).
  - Blades with adaptive twist angle by means of distributed actuator material (individual blade control).

A market study performed by a well-respected consultancy around 1992 projected:

- Engine-mount vibration isolators and transmission pressure plates to emerge within a year or two.
- Large number of further applications to emerge in the 3 to 10 year time frame.
- Smart materials to provide 'arms' and 'fingers' mechanically simpler and thus more reliable than conventional electromagnetic devices.
- An initial $100 to $150 million worldwide market to turn into a billion-dollar market by 1995.
- Vibration control for aerospace and automotive as well as other applications, a multi-billion-dollar market to be dominated by smart materials.

- Advanced composites market to increase from $2–$3 billion to $20 billion or more by the end of the 20th century with smart composites easily taking a share of up to 50 %, with the aerospace market being the driving force followed by automotive, civil engineering and leisure market.

What happened to all this? When looking around these days we can hardly see any of these predictions to have come true. Was this mainly a fake? Or have we been too impatient?

A look at the development of other technologies shows that two decades are not sufficient to make a conclusion. Traditionally the incubation time of a technology has more to be seen within the range of half a century. Hence making a judgement today on smart technologies' progress in aerospace has to be seen more in terms of a 'mid-term review'. A major reason for this slow progress is the long life cycle of products in aerospace. Most of the fixed wing aircraft flying today were built around two decades ago with design principles developed over the past half century. These principles still govern our aircraft design today and as such most of the aircraft flying for two decades and being built today. The complexity our aircraft have achieved and the airworthiness regulations they have to meet do not give them a great margin to adopt smart technologies in the short term. However, the impact of smart technologies could be further enhanced if an aircraft's structural design and life cycle process were better understood and the economic rewards could be better grasped.

Within the following sections a trial is made on how to spot the aerospace-related view on smart technologies. It starts by looking at what achievements have been obtained over the past two decades, followed by methods on how to get those achievements integrated into existing and future aerospace structures and finally looking at how these prospects may mature in the longer term.

6.2 THE PAST TWO DECADES

The variety of research done over the past two decades in developing smart technologies for aerospace structures is well reflected in the numerous publications that can be found among different sources. Different overview papers related to aerospace appeared in *Smart Materials and Structures* published by the Institute of Physics since 1992 and the *International Journal on Intelligent Material Systems and Structures* published by SAGE Publications since 1990. Another important source in that regard is the *AIAA-Journal*. *Smart Materials and Structures* has published a number of aerospace-related special issues on space [2, 3] and rotorcraft [4, 5]. A number of overview articles do also report on developments related to rotorcraft or other aerospace issues [6–9]. There is also a book on health monitoring of aircraft which largely encompasses the smart technologies aspect [10]. A good overview of smart technologies developments in various fields including aerospace has been given in the book *Adaptronics and Smart Structures* edited by H. Janocha [11], of which a second edition is due to appear shortly and where a comprehensive summary of aerospace-related developments will be provided [12].

Another significant source of information is the various conferences and symposia being held around the world. A 'product' of smart technologies research in aerospace is the *International Conference on Adaptive Structures (ICAST)* held annually since 1991. The largest forum with a high aerospace content is possibly SPIE's annual *International Symposium on Smart Structures and Materials* [13] followed by events such as CanSmart [14]. SPIE also hosts a variety of further conferences such as those in Australia and India. There are some further events of a more national character such as AIAA's Structural Dynamics Conference in the USA or the Adaptronic Congress in Germany where different aspects of smart technologies in aerospace are discussed as well. Further aerospace-related conferences such as the *International Conference on Aeronautical Structures (ICAS)* [15] have also well accommodated the smart structures aspect. Structural health monitoring (SHM) which has become highly relevant with aircraft is mostly discussed separately such as at SPIE's annual *Nondestructive Evaluation for Health Monitoring and Diagnostics Symposium* [13], the *International Workshop on SHM* [16] and the *European Workshop on SHM* [17], or at other aerospace-related events where this topic has turned out to be significant.

When trying to structure the different developments made in smart technologies for aerospace structures the following categories can be identified:

- SHM,
- shape control and active flow,
- damping of vibration and noise,
- smart skins, and
- systems.

The following subsections will try to summarise most of the achievements in these different areas so that a sufficiently clear overview may be given.

6.2.1   SHM

Whether SHM has been triggered by smart technologies or not can remain an issue for an endless debate. Principally structural health became an issue when damage tolerance principles were introduced more than half a century ago. Very shortly after the first accidents occurred with the Comet aircraft in 1954 it became obvious that aircraft would have to be equipped with additional sensors. This was highly limited to monitoring parameters characterising the operational loads such as accelerations or strains and taking advantage of the analytic and prognostic algorithms and knowledge having been generated at the time. The outcome of roughly four decades of R&D work has resulted in some fighter aircraft now possessing a set of either conventional electrical strain gauges or recordings from a limited number of already existing sensors to record flight parameters, which allow us to construct and trace the operational loads spectrum the aircraft have flown in and to evaluate their structural life consumption as well as their potential residual life. The most advanced aircraft in that regard is possibly the Eurofighter Typhoon [10].

Helicopters faced a similar problem some time later but thus more abrupt. The near-to-constant amplitude loading sequence of helicopter components and the resulting very short crack propagation life led to the loss of various helicopters at very short notice and triggered the need for a system to recognise structural anomalies at a very early stage. This resulted in the development and introduce Health and Usage Monitoring Systems (HUMS) which now are mandatory in any new helicopter [18].

If smart technologies did trigger SHM then this definitely emerged with carbon fibre reinforced composites (CFRPs) and their ability to integrate sensors and actuators, specifically if they could be miniaturised. The major driver for this was the likelihood of barely visible impact damage (BVID) appearing, which could result from unmonitored accidental damage and would not be seen from the outside by conventional visual inspection. This fear partially results from a metallic materials design mindset and has so far not proven to be relevant as long as composite material structures have been designed for a safe life. However, SHM may become increasingly relevant the more damage tolerance potential is recoverable from this type of structure.

SHM applications therefore moved back again to metallic structures where monitoring had been much more relevant, mainly driven by the increasing number of ageing aircraft [19] as well as the enhancement of damage tolerance principles [20]. Schmidt H.-J. and colleagues [20] showed in a series of papers that the introduction of SHM into conventional damage-tolerant

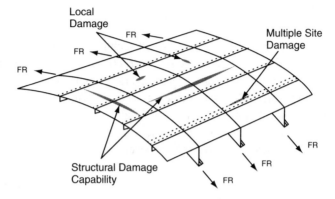

Figure 6.1 Assumed crack propagation scenario for the design of aircraft fuselage structures [20]

design could reduce the weight of a component by as much as 20 %. The idea is based on the fact that many airframe structures such as the one shown in Figure 6.1 can only be inspected from the outside. Since the crack propagation life of the stringer would be short compared with the crack propagation life of the skin panel (Figure 6.2) the stringer (or frame) would have to be assumed to be broken. However, if the stringer or frame could be monitored the assumption of a broken stringer or frame could be dropped and a crack propagation behaviour would be obtained as shown in the right hand scenario of Figure 6.2. Since this would result in longer inspection intervals,

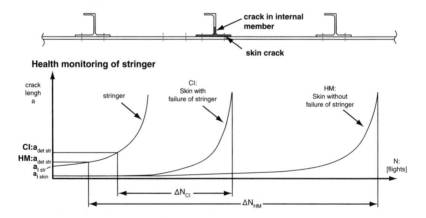

Figure 6.2 Comparison of fuselage skin crack propagation life for an unmonitored and monitored frame respectively [20]

this could also be traded for higher allowable stresses and lower mass and weight in case the inspection intervals cannot be altered.

These findings have driven the major aircraft manufacturers to put further emphasis on the SHM topic. Their motivation so far has been to:

- Monitor load exceedances and predict their impact on the aircraft structure's condition.

- Look at areas which are prone to damage but difficult and thus costly to access and as such would be useful to be monitored in the sense of automating an inspection procedure.

- Screen technologies which exist and which might be beyond the state of the art.

The first motivation has resulted in providing flight-parameter-based hard landing monitoring systems such as there principally shown in Figure 6.3 and now intended to be offered for aircraft such as the Boeing 787. These systems are based on monitoring flight parameters which are then fed into a neural network that allows the loading pattern and as such the stresses and strains in the landing gear and the aircraft's structure to be determined, provided a digital model of the structure is available.

Aircraft can be subject to a variety of non-standard loads such as collisions with ground vehicles around doors and specifically cargo doors, which can possibly even endanger their closure mechanisms, or enhanced humidity around galleys and toilets. Figure 6.4 gives some idea of where smart sensing systems may be placed first such that an enhanced sensing effort at these locations could be automated and where sensors could be validated in the first place to the benefit of the operator.

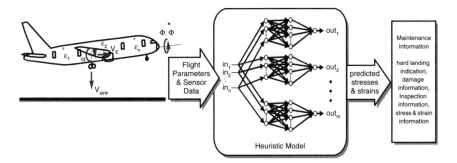

Figure 6.3 Hard landing data collection using flight parameters (Courtesy of Boeing) [21]

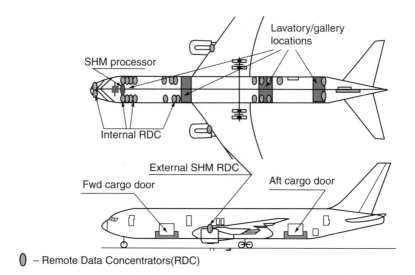

Figure 6.4 Potential locations for initial SHM on an aircraft

Finally there is a large evaluation currently ongoing in assessing different types of load and damage monitoring sensors and sensor systems. The most popular of those include:

- Optical fibre Bragg gratings for monitoring loads, humidity or pressure and partially also damage at various locations [22, 23].
- Piezoelectric sensors such as Smart Layers [24] or piezofibre-based sensors [25].
- Eddy-current-based foils [26] that can be integrated into metallic structures where a magnetic field is generated via a number of conductive metallic windings and recorded by another number of windings, all configured as shape-field sensors on a carrier.
- Microelectronic mechanical systems (MEMS) for monitoring environmental degradation such as corrosion (i.e. [27]).
- Comparative vacuum monitoring from structural monitoring systems [28].

All of these sensor types and monitoring technologies are currently part of extensive in-service trials on aircraft.

6.2.2 Shape Control and Active Flow

Shape control and thus actively influencing the flow of aerodynamic profiles is possibly the ultimate aim in aerospace and what smart structures technology has been expected to provide. As a consequence, different programmes were launched, among which the Smart Wing Program [29] run in the USA in the early 1990s can be considered as the pilot study. It was quickly learned that neither actuators possibly machined from bulk shape memory material nor piezoelectric wafers attached to an aerodynamic profile's surface would meet any requirements, mainly for reasons of the energy required or the additional weight imposed. It was therefore very quickly concluded that any solution for adaptive wings would only hold if they were to stay within the principles of conventional solid state actuators and that these actuators could be made from an actuation material such as a piezoelectric. The next step therefore resulted in a much more conventional mechanical solution actuated by piezoelectric travelling wave ultrasonic motors as shown in Figure 6.5 [30]. This solution was then built into a 30 % downscaled model of an uninhabited combat air vehicle (UCAV) of which the control surfaces consisted of a conventional flap solution (left) and a segmented continuously deforming smart solution (right) as shown in Figure 6.6. In a wind tunnel test at Mach 0.8 wind speed and 300 psf dynamic pressure the smart solution showed a 17 % improvement in rolling moment coefficients at 15 degrees of control surface deflection when compared with the conventional flap solution [31].

The Smart Wing Program together with various other programmes related to adaptive aerodynamic profiles was placed under the NASA Morphing Wing Program [32] which emphasised a strategic direction and the need to place different disciplines under one umbrella. Another major DARPA project was the Compact Hybrid Actuator Program (CHAP) from which SAMPSON [33] emerged as a key project. SAMPSON dealt with an adaptive air inlet duct of the engine of an F-15 aircraft that allowed the cowl to rotate, the lip to deflect, the air intake wall to deflect and the lip to blunt. This was achieved mainly by using SMA actuators configured as either a rod

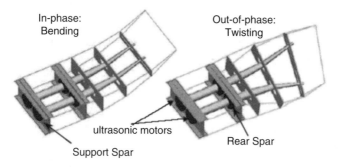

Figure 6.5 Actuation principle for a hinge-less control surface segment based on an eccentuator and ultrasonic motor [30]

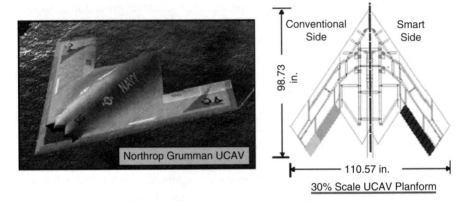

Figure 6.6   Northrop Grumman's Smart Wing Model Phase II [30]

of 60 SMA wires, or wires placed in a flex skin panel. Wind tunnel tests were performed to determine the gains in noise reduction. Other solutions for engine noise reduction include chevrons around the engine exhaust which are activated by SMA actuators in accordance with an aircraft's flight level position.

Further ideas which deal with the adaptive control of fluid flow around aerodynamic profiles to increase mainly lift have been pursued either by adaptively creating a bump and thus moving the point of supersonic fluid flow separation further down the profile or in delaying turbulent boundary layer separation by superimposing forced oscillations on the mean flow being on the verge of separation. This latter approach may be achieved by synthetic jets or any other mechanical devices including phased plasmas.

Problems and issues arising with fixed wing aircraft can of course also very easily be found with rotary wing aircraft. To control flow on this kind of profile has resulted in solutions such as synthetic jets or active flipperons as well [34, 35]. The jets operate similar to the piezoelectric jets used on inkjet printers, while the flipperons have been configured as sandwiched multilayered PVDF (Poly VinyliDene Fluoride) beams or as PZT (Lead Zirconate Totanate, piezoelectric ceramic material) single crystals. Generally significant improvements in drag reduction (12 %), lift (10 %) and angle of attack (20 %) were observed.

Most of the initial smart-technologies-related work on shape control originated from the shape control of large space antennas, mirrors and solar panels using piezoelectric wafer actuators. This work has been currently enhanced by using piezofibre actuators which are more flexible and can be better integrated into complex-shaped composite structures.

Finally it should be mentioned that many of the ideas mentioned above have been applied to missiles and have been possibly further enhanced, but due to the classified nature of that area not much has been disseminated into the public domain.

6.2.3 Damping of Vibration and Noise

The high-frequency operability of piezoelectric actuators very quickly made them attractive to solve vibration and as such aeroelastic and flutter problems on aerodynamic profiles. A remarkable study was performed at NASA Langley together with the Royal Australian Air Force on the fin of the F-18 aircraft to alleviate fin buffet problems [36]. Piezoelectric wafers were again bonded to well-selected locations of the fin, initially to a downscaled model that was wind tunnel tested and then to a real fin that was evaluated in a flight test and where improvements could be observed.

The issue of buffeting was also widely analysed within an EADS/DLR programme [37, 38]. Different solutions as shown in Figure 6.7 were considered, of which three were finally selected for further in-depth analysis. The first one resulted in a large slightly downscaled fin that was equipped with 2410 piezoelectric wafers and run on a vibration test for which some of the results are given in Figure 6.8. The main conclusion from that test was that the first bending and torsion mode could be well reduced and that the piezoelectric wafers turned out to be less of a weight penalty compared with the additional amplifier equipment that would have had to be taken onboard the aircraft.

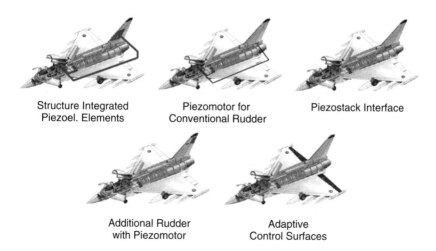

Structure Integrated
Piezoel. Elements

Piezomotor for
Conventional Rudder

Piezostack Interface

Additional Rudder
with Piezomotor

Adaptive
Control Surfaces

Figure 6.7   Different options for buffet alleviation using smart technologies [37, 38]

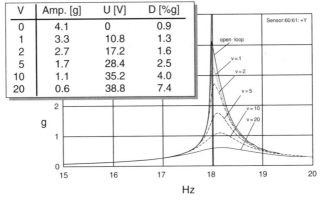

V	Amp. [g]	U [V]	D [%g]
0	4.1	0	0.9
1	3.3	10.8	1.3
2	2.7	17.2	1.6
5	1.7	28.4	2.5
10	1.1	35.2	4.0
20	0.6	38.8	7.4

Figure 6.8   Fin box with adapted piezoelectric actuators [37, 38]

The other option in actuating an additional rudder through a piezoelectric motor has been shown numerically to work quite efficiently, while the option with the adaptive control surfaces would virtually work but mainly suffered from the lack of actuators providing either sufficient stroke or force [39].

Vibration and as such buffeting problems could also be solved through a change in a structure's stiffness. This has been achieved in the ADAPT project where SMA wires were integrated into the glass fibre composite skin of an aerodynamic profile of about 0.5 metres in size (Figure 6.9) [40]. Simply heating the SMA wires into the austenitic condition showed that the tip deflection in a simple vibration test could be cut by half and the eigenfrequency of the first bending mode moved.

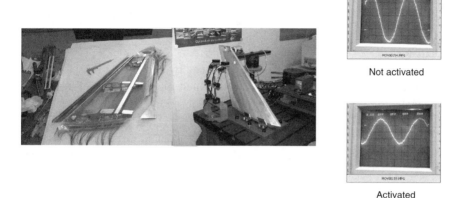

Not activated

Activated

Figure 6.9 Fin with SMA-reinforced composite skins [40]

For rotorcraft the issue of vibration damping and resulting noise from the rotor blades has been tackled mainly by two solutions: directionally attached piezoelectric (DAP) actuators [41] or flaps at the leading and trailing edge of a rotor blade [42]. While the former is a solution mainly applicable to very small rotor blades in the size of a model airplane, the latter has been realised and tested at full scale [43]. Figure 6.10 shows the principle, which consists

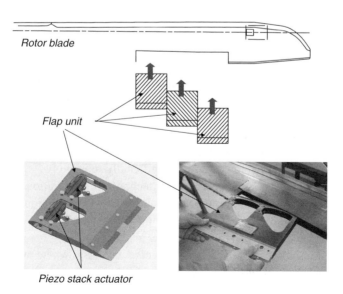

Figure 6.10 Piezo stack actuated flap units for helicopter rotor blades [42]

of flap units that are integrated into the rotor blade. The flaps are actuated by piezoelectric stack actuators and a leverage system that allows the required stroke to be realised. Further damping and actuation mechanisms of a similar nature have been suggested on the basis of SMA or magneto-rheological actuators.

Cabin noise in helicopters has been reduced by decoupling the cabin from the gear box by integrating smart struts which mainly consist of a stacked piezoelectric actuator that counteracts the vibrations generated by the gear box [44].

With respect to space vehicles, piezoelectric actuated struts were developed in the late 1980s and were considered to actively dampen vibrations of large space vehicles. A spectacular success for any smart structures application was shown with the articulating fold mirror of the Hubble Space Telescope where piezoelectric actuators allowed the jitter produced by an adjacent cooler to be compensated. Solutions of that nature were further explored as a means of vibration isolation for other types of equipment transported on spacecraft [45]. Noise impacts on payloads have been dealt with as well, where acoustic foam combined with actuation material such as PVDF has been under examination [46].

Finally it should also be mentioned that active mounts have helped to meet military specifications for conventional off-the-shelf electronic components and could generate a cost saving potential in the defence sector.

6.2.4 Smart Skins

The topic of smart skins has been generally included in the discussion of smart structures in aerospace. It includes the integration of various types of antennas into the air vehicle's structure, thus giving the antennas a load-bearing and communication multifunction and further removing the antenna as an aerodynamic obstacle from the aircraft's surface. Furthermore, antennas can be clustered in antenna apertures which will allow the number of antennas as well as their weight and cost to be significantly reduced [47]. Solutions in that regard can be seen with phased array technology. Integration of antennas will further require steering of the antenna's beam where solutions have been proposed as a sandwiched structure containing a metal plate activated by piezoelectric actuators such as a Rainbow actuator [48].

Smart skins also include radar absorbent coatings where a variety of compositions have been developed or are further under development. This also includes smart microwave windows which allow microwave reflectivity to be changed through a change in electrical dc potential. An adaptive radar-absorbing structure based on the topology of a Salisbury screen combined with an active frequency selective surface controlled by PIN diodes which allows for superior reflectivity–bandwidth has been reported in [49].

MEMS in their second generation have now become much more feasible for mass production. As such they have increasingly been considered to be integrated into structural skins allowing fluid shear stresses to be measured on aerodynamic profiles [50]. Antennas based on RF-MEMS integrated into a structure's surface are another interesting area within the topics of satellite communication and space-based radars [51].

6.2.5 Systems

Although a smart structure should be a smart system within itself, the solutions and ideas presented above have been mainly add-ons to a conventional engineering system such as an aircraft. The danger with that approach has to be seen with the degree of freedom in design and the design principles used. The more complex an engineering system such as an aircraft becomes, the more the design principles applied are 'frozen' and the less a freedom in design applies. This limited freedom in design may explain why a variety of the solutions described previously have stayed at a level of feasibility demonstration rather than moving forward into real applications. If smart technologies therefore should move into true application and thus production one needs to explore the fields where the freedom of design is still wide.

An area in aerospace engineering where a relatively large freedom of design is still given is micro aerial vehicles (MAVs). MAVs have seen incredible progress, specifically with the integration of smart technologies. This trend possibly started with the integration of the directionally attached piezoelectric (DAP) actuators on the rotor blades of a miniaturised helicopter [41] followed by piezoelectric benders as stabilators of a rotary wing MAV [52] (Figure 6.11). Antagonistic filaments of SMA wires have been shown to work as actuators for roll control of an MAV that has been generated through a change in the wing angle pitch [53] (Figure 6.12).

Electro-active polymers (EAPs) are another type of material that has been used as an actuation mechanism and realised for a flapping wing MAV [54]. MAVs have also been explored down to the micro scale with a gross weight of 7.5 grams, but with flight durations currently just in the range of seconds than minutes. MEMS technology has been mainly applied for the manufacture of this type of veined wing structure derived mainly from the structure observed on an insect's wing [55] (Figure 6.13).

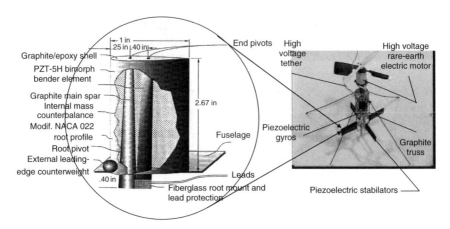

Figure 6.11 Piezoelectric actuated control surfaces on rotary wing MAV [52]

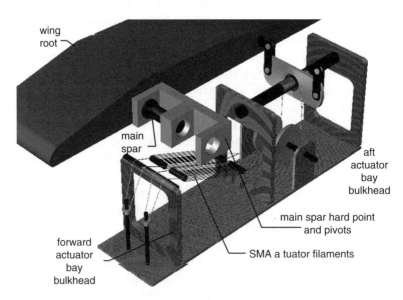

Figure 6.12 Antagonistic SMA actuation system for wing pitch [53]

Further enhancement shown in Figure 6.14 has been achieved by developing an airflow control mechanism for these insect-like MEMS-built wings which consists of an array of valves where each valve consists of a tethered cap, such that it can be actuated individually.

Finally, as with aircraft and MAVs, there is a similar trend observed between conventional satellites and micro-satellites, where the latter still

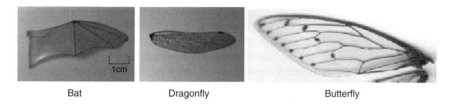

Bat   Dragonfly   Butterfly

Figure 6.13   MEMS-based veined wing structures for MAVs [55]

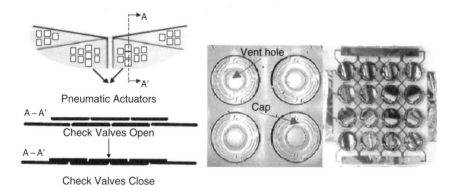

Figure 6.14   Parylene-valved actuator skins for MAV wings [55]

possess a much larger freedom of design and where a variety of MEMS technology is integrated for communication, docking or actuation purposes.

6.3   ADDED VALUE TO THE SYSTEM

New technologies such as the smart technologies considered here will only be introduced if they provide added value to the product or system. This added value is often difficult to determine because it is determined more from a marketing, product and process planning rather point of view than from an engineering one. However, some of this marketing and process knowledge is required to understand completely the definition of a successful product.

One 'tool' to help possibly in that regard is target costing. Target costing is a methodology (or possibly even a philosophy) that allows us to transfer market needs and thus customer value into hard fact product features appreciated by the customer. Seidenschwarz W. *et al.* [56] mechanistically structured the target costing process in the form of a tool box, which comes close to the language of an engineer. Figure 6.15 shows this 'tool box' with all its different 'tools'.

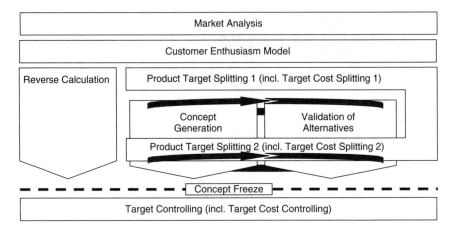

Figure 6.15 The target costing tool box [56]

The first 'tool' to consider is market analysis. This is mainly related to deriving the customer's requirements with respect to the product considered. These requirements can be determined through interviews, questionnaires or even specific 'customer clinics', to name just a few. This, however, also requires engineering judgement in terms of what requirements have to be specifically achieved (basics) and which requirements could be established as a result of new technologies and enthuse the customer, basically following the principle 'What is going to make our product different to the one from our competitor?' An essential requirement which has to come out of that analysis is the price a customer is willing to pay for the requirements to be achieved, i.e. the product.

This now leads on to the second 'tool' which is the customer enthusiasm model, schematically shown in Figure 6.16. This model mainly allocates the different requirements to three different categories termed 'basic', 'performance' and 'enthusiasm'.

'Basic' includes all requirements mainly not expressed by the customer and which will hardly be appreciated irrespective of their completeness. However, missing one of these requirements can be the reason for complete product failure. 'Performance' includes all requirements that are usually expressed by the customer and which the customer is fairly well able to value to some degree. This also includes the price the customer is willing to pay for the product. 'Enthusiasm' finally includes those requirements where the customer values the requirement much higher than the cost of its achievement. These are the requirements which often the product manufacturer also and thus engineers bring in terms of new technology to be implemented.

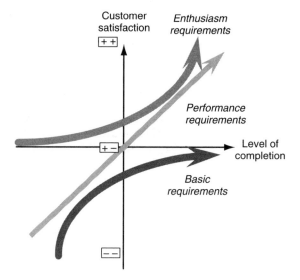

Figure 6.16   The customer enthusiasm model

Knowing the price the customer is willing to pay for the product, the allowable cost for it to be produced can now be determined along a reverse calculation. This is done by gradually slicing off all the cost elements an engineer cannot influence from the price the customer has expressed. This includes elements such as VAT, profits, overheads, administration, financial obligations, insurances, etc., and results in the allowable production target cost.

This allowable production target cost now needs to be allocated to the different components of the product. This happens a product target splitting process where the principles of quality function deployment (QFD) are applied and which is schematically shown in Figure 6.17.

Let us assume that the customer has expressed eight different requirements and has been asked to give each one a weight in per cent such that the sum of these weights totals 100%. Let us further assume that our product consists of five components which are given on the top of the QFD matrix. An engineering or product design team will now explore how much each requirement contributes to each component and will write the number decided in each box of the QFD matrix. For this, there will again be 100% to be distributed per requirement. Hence requirement 1 contributes 5% to component 1, 30% to component 2, 0% to component 3, 50% to component 4 and 15% to component 5, which all sums to 100% in total. The same is done for the other seven remaining requirements.

		Component					
	Weight [%]	1	2	3	4	5	Total
Requirement 1	7	5	30	0	50	15	100
Requirement 2	10	0	25	0	75	0	100
Requirement 3	15	0	0	100	0	0	100
Requirement 4	12	10	5	30	0	55	100
Requirement 5	5	30	0	35	35	0	100
Requirement 6	20	20	25	10	25	20	100
Requirement 7	22	20	0	0	80	0	100
Requirement 8	9	20	20	20	20	20	100
Total [%]	100						

Figure 6.17 QFD matrix for product target splitting

In a next step the weight of each requirement is multiplied by the percentage the requirement contributes to the component. Thus 7 % is multiplied by 5 % and leads to 0.0035 or 0.35 %, 7 % is multiplied by 30 % and leads to 2.1 % etc. for requirement 1, 10 % is multiplied by 0 % to obtain 0 %, 10 % is multiplied by 25 % to obtain 2.5 % etc. for requirement 2 etc. until all the components' related numbers in the QFD matrix are multiplied by the weights of the respective requirements. This results in the matrix given in Figure 6.18.

If the percentages in each column are added up we obtain the percentage that the respective component would be allowed to cost. For component

		Component				
	Weight [%]	1	2	3	4	5
Requirement 1	7	0.35	2.1	0	3.5	1.05
Requirement 2	10	0	2.5	0	7.5	0
Requirement 3	15	0	0	15	0	0
Requirement 4	12	1.2	0.6	3.6	0	6.6
Requirement 5	5	1.5	0	1.75	1.75	0
Requirement 6	20	4	5	2	5	4
Requirement 7	22	4.4	0	0	17.6	0
Requirement 8	9	1.8	1.8	1.8	1.8	1.8
Total[%]	100	13.25	12	24.15	37.15	13.45

Figure 6.18 Resulting QFD matrix after multiplication

1 this is 13.25 %, for component 2 12 %, etc. So if the target production cost for the product € 100 then € 13.25 would be allowed for the production of component 1, € 12 for the production of component 2, etc.

The product target splitting which has been shown as a one-step approach here can also be done in two steps, which is specifically advisable when the requirements set are less of a technical but more of a soft psychological nature. In that case the soft requirements are first converted to features of a more technical nature by applying the same procedure as the one described above. In the second step (product target splitting 2) the features with their resulting weights are then converted to take the role of the requirements and the procedure is again the same as described above.

The flavour of this procedure is that one obtains the value that the customer appreciates for each component of the product. This percentage (or target cost) can now be compared with the real cost for the component to be produced. In case the target cost is higher than the real cost then the component is undervalued by the manufacturer and could be budgeted higher, while the opposite case means that cost reductions in terms of redesign or a re-evaluation of the requirements would be required.

Within that cost assessment process the implementation of smart technologies has to be discussed. Smart technologies can contribute to a product if they either help to reduce the cost of a costly component, or produce additional value to the customer, where the additional value has to be at least as much as the smart technology may cost. The conclusion here is that smart technologies may only be successfully integrated into a product if the product requirements as well as the product's cost splitting process are well known. Obtaining that information for an engineer sounds hard, because it may require reliable links to marketing and finance departments. These links may certainly be of significant help. However, with some understanding the marketing and financial information may even be generated and validated by an engineer only without any external help, at least to a first valuable approach.

That an understanding of product processes is essential to identify the potential of smart technologies can also be seen in another context. SHM is a technology very much discussed with respect to aircraft structural integrity. SHM may help to:

- Speed up the inspection process and thus increase aircraft operability.
- Automate the inspection process and thus free people from less efficient inspection work.

- Enhance the frequency of inspections to increase the operational life of components without compromising safety.

- Expand the idea of damage-tolerant design and allow components to be designed with lighter weight.

Just take the first of these four drivers and imagine how much one day of non-availability of an aircraft of the category of an Airbus A380 (retail price around €200 million) would cost the operator: €750 000. Saving maintenance time and thus enhancing operability will therefore be paramount to the aircraft operator when compared with automating the inspection process or purchasing additional inspection equipment. This prioritisation results from the high asset value an aircraft has and may explain why SHM is currently discussed more with that category of product than with a comparatively lower asset value product such as an automobile.

To find out if SHM may help to speed up the inspection process requires the maintenance process to be analysed in its entirety. It needs to be understood which maintenance actions are run in series, which in parallel, and which process elements are along the critical path of the maintenance process. The critical path then has to be analysed with respect to the maintenance actions that are relevant to structural inspection. If there are no structure-related inspection actions along the critical path, then SHM will not help to enhance aircraft operability. However, if there are, then the simulation process in finding out the SHM potentials will move ahead. A second step would therefore be to analyse all components to be inspected along the critical path of the maintenance process with regard to a possible SHM solution, i.e. what monitoring principle, sensor type, data acquisition method, etc., would have to be used. The cost of this solution in terms of total life cycle cost and process times with regard to installation and maintenance of the SHM solution, as well as the process times for the inspection process itself, would have to be determined. This information would then have to be fed back into the maintenance process simulation which would now demonstrate the time that the maintenance process would have been reduced and aircraft operability increased. It would further show if the maintenance action considered would still be located on the critical path of the maintenance process or if the critical path would have moved. If the latter happened, then the new critical path would again have to be traced with respect to elements of structural inspection relevance which may then have to be analysed in the same way as described before.

Analysing engineering systems and processes in the more holistic way described here indicates how added value to a product can be understood and how smart technology solutions can be derived that will then find their way into successful applications. Simulating these processes is therefore an essential prerequisite to get smart technologies into application.

6.4   POTENTIAL FOR THE FUTURE

So far a lot of research has been performed in aerospace-related smart structures technology but only very little has gone into true application. Is this because this technology has so little potential and may have more intellectual than commercial rewards? This question may have different answers, as given below:

1. Every technology takes about half a century to achieve its breakthrough from the very first beginning when the idea was born. This sounds very depressing in a time when expectations of 'time to market' increase continuously. However, it is not the time of physical development that drives the development process but the time of understanding. Unfortunately the latter can hardly be enhanced since it is mainly implemented by human nature. To recall a practical example in aviation already partially mentioned above: the need for a loads monitoring system in aircraft was recognised after the Comet aircraft crashes in 1954. The most sophisticated solution which has come out today in that regard is the Eurofighter Typhoon with 16 strain gauges (in the UK and Spanish versions). This indicates the progress achieved over half a century irrespective of all the technology being available.

2. We have been focusing too much on a specific technology we have developed instead of the broader picture of what society and products need. Product development therefore needs to become much more interdisciplinary to understand where we really stand. This, however, was never truly promoted in the past and may have to be done so in the future.

3. We have had too many duplications of smart technology research in the past and have thus not used our resources effectively. This again resulted from competing researchers and research organisations as well as from the lack of knowledge of funding organisations or the reviewers these organisations consulted. Although the research community meets at numerous conferences, symposia or workshops, this hardly allows the duplication of research to be avoided because research reported is a 'post artefact' that can only be recognised and not be changed by any other means.

What is lacking in the implementation of smart technologies is a more strategic approach. This could be provided by a more holistic view including simulation of design, product generation and product life cycle processes, which would enlighten the understanding of product, market and society needs as well as the value the society would appreciate. Furthermore the research community would require more strategic guidance. This could be achieved by funding organisations staging strategic seminars that would map the research skills available and allocate the resources in accordance with these skills so that duplication in research would be minimised. This would of course have to remove some of the competitive nature of research, but it could possibly be a much more efficient way of using the limited resources with respect to time, quality and cost of research performed.

When analysing the potential of smart technologies for the future one might say that they have still not been adequately explored. We are still mainly at the stage where smart technologies emerged around two decades ago, so our projections for this potential are still positive. However, our projections for the future are not built on very solid grounds, which may be essential at this stage to indicate clearly which way to move in the future. From our feelings we can say that SHM or MAVs may be the promising smart technology of future applications, but emotions have hardly been suitable grounds for engineering to be built upon.

6.5   A REFLECTIVE SUMMARY WITH CONCLUSIONS

When looking back at what has been explored in the context of smart technologies in aerospace over approximately the last two decades, one can conclude that the outcome deserves respect. Various aerospace-related problems from the past have been used as a benchmark and for reiteration to see how far smart technologies may be able to solve them. Many useful solutions have been produced which may currently rest in some drawers, shelves and in folders, mainly because the people in charge of them do not possibly know how to proceed from there.

Smart technologies have also brought together communities from different disciplines mainly in the natural sciences, which used to operate more in isolation during the past. Interacting as a structural engineer with materials scientists, control experts, electronics specialists, chemists, biologists, etc., has been an unbelievable enrichment for this author towards a broader and thus holistic understanding. Such processes of interdisciplinary mergers take a long time and must also be considered as a cultural problem. In view of achieving the broader holistic view of a product, this trend in mergers is vital and thus has to be pursued by all means.

Smart technologies have produced an increasing number of scientific output in terms of journal and conference papers. The two major journals in the area, *Smart Materials and Structures* and *Intelligent Materials Systems and Structures*, have significantly increased in size and have achieved impact factors of above 1.6 and above 1.0 respectively. SHM has emerged as a separate field of science with at least one international conference held annually. Two journals, *Structural Health Monitoring* and *Structural Control and Monitoring*, have either fully or partially devoted their focus to this topic.

As such the research output in smart technologies has been booming and prompts the question of where this should all lead to in the end. Due to the interdisciplinary culture this has created, this trend should not stop but needs to find its true applications and this may only occur when the appropriate strategic tools are provided. Some suggestions have been made here on target costing and life cycle process simulation, but it is far too early to say if there might be the right tools or not. This needs to be explored and discussed with possibly further interesting assessment procedures emerging once the discussion has been expanded into further areas of science such as marketing or economics. This would be in the nature of smart technologies and would help to bring this field to the point where it is intended to go.

REFERENCES

[1] Breitbach E.J., 1991: *Research Status on Adaptive Structures in Europe*; Proc. of 2nd Joint Japan/US Conf. on Adaptive Structures; Technomic Science Publisher

[2] Bronowicki A.J., A. Das and B.K. Wada (Eds.), 1999: Special Issue on *Smart Structures for Space*; Smart Mater. Struct., **8**, No. 6

[3] Böhringer K.F. (Ed.), 2001: Special Issue on *Space Applications for MEMS*; Smart Mater. Struct., **10**, No. 6

[4] Chopra I. (Ed.), 1996: Special Issue on *Application of Smart Structures Technology to Rotorcraft Systems*; Smart Mater. Struct., **5**, No. 1

[5] Gandhi F. (Ed.), 2001: Special Issue on *Rotorcraft Applications*; Smart Mater. Struct., **10**, No. 1

[6] McGowan A.-M.R., A.E. Washburn, L.G. Horta, R.G. Bryant, D.E. Cox, E.J. Siochi, S.L. Padula and N.M. Holloway, 2002: *Recent Results from NASA's Morphing Project*; Proc. of SPIE Vol. 4698, pp. 97–111

[7] Garcia E., 2002: *Smart Structures and Actuators: Past, Present, and Future*; Proc. of SPIE Vol. 4698, pp. 1–12

[8] Loewy R.G., 1997: *Recent Developments in Smart Structures with Aeronautical Applications*; Smart Mater. Struct., **6**, No. 5, pp. R11–R42

[9] Giurgiutiu V., 2000: *Review of Smart-Materials Actuation Solutions for Aeroelastic and Vibration Control*; J. Int. Mater. Syst. Struct., **11**, pp. 525–544

[10] Staszewski W.J., C. Boller and G.R. Tomlinson, 2003: *Health Monitoring of Aerospace Structures*; John Wiley & Sons, Ltd

[11] Janocha H. (Ed.), 1999: *Adaptronics and Smart Structures*; Springer-Verlag

[12] Boller C., 2006: *Adaptronic Systems in Aeronautics and Space Travel*; in Janocha H. (Ed.): *Adaptronics and Smart Structures*; 2nd Edition, Springer-Verlag

[13] http://www.spie.org/info/ss-nde

[14] http://www.cansamrt.com

[15] http://www.icas.org

[16] http://structure.stanford.edu/workshop

[17] http://www.shm-europe.net

[18] http://aar400.tc.faa.gov/Programs/agingaircraft/rotorcraft/index.htm

[19] Boller C., 2005: *Process Chain Elements for Health Management of Engineering Structures*; Proc. of the 5th Int. Workshop on Structural Health Monitoring; DEStech Publications Inc.

[20] Schmidt H.-J. and B. Schmidt-Brandecker, 2001: *Structure Design and Maintenance Benefits from Health Monitoring Systems*; in Chang F.-K. (Ed.): *Structural Health Monitoring*; CRC Press, pp. 80–101

[21] Boller C., A. Akdeniz, H. Speckmann and M. Buderath, 2006: *Structural Health Monitoring Approaches to Fixed Wing Aircraft*; Manuscript submitted to Smart Mater. Struct.

[22] Read I.J. and P.D. Foote, 2001: *Sea and flight trials of optical fibre Bragg grating strain sensing system*; Smart Mater. Struct., **10**, pp. 1085–1094

[23] Betz D., 2004: *Acousto-ultrasonic sensing using fibre Bragg gratings*; PhD thesis, Department of Mechanical Engineering, Sheffield University, UK

[24] http://www.acellent.com

[25] http://www.smart-material.com

[26] Washabaugh A., V. Zilberstein, D. Schlicker, I. Shay, D. Grundy and N. Goldfine, 2002: *Shaped-Field Eddy-Current Sensors and Arrays*; Proc. of SPIE Vol. 4702, pp. 63–75

[27] http://www.analatom.com

[28] http://www.smsystems.co.au

[29] Kudva J.N. et al., 1999: *Overview of the DARPA/AFRL/NASA Smart Wing Program*; Proc. of SPIE Vol. 3674, pp. 230–236

[30] Kudva J.N., B. Sanders, J. Pinkerton-Florance and E. Garcia, 2001: *Overview of the DARPA/AFRL/NASA Smart Wing Phase 2 Program*; Proc. of SPIE Vol. 4332, pp. 383–389

[31] Scherer L.B., C.A. Martin, B. Sanders, M. West, J. Florance, C. Wieseman, A. Burner and G. Fleming, 2002: *DARPA/AFRL Smart Wing Phase 2 Wind Tunnel Test Results*; Proc. of SPIE Vol. 4698, pp. 64–75

[32] McGowan A.-M.R., A.E. Washburn, L.G. Horta, R.G. Bryant, D.E. Cox, E.J. Siochi, S.L. Padula and N.M. Holloway, 2002: *Recent Results from NASA's Morphing Project*; Proc. of SPIE Vol. 4698, pp. 97–111

[33] Pitt D.M., J.P. Dunne and E.V. White, 2002: *SAMPSON smart inlet design overview and wind tunnel test; Part 1 – Design overview; Part II – Wind tunnel test*; Proc. of SPIE Vol. 4698, pp. 13–36

[34] Jacot D., T. Calkins and J. Smith, 2002: *Boeing Active Flow Control System (BAFCS)-III*; Proc. of SPIE Vol. 4698, pp. 76–84

[35] Calkins F.T. and D.J. Clingman, 2002: *Vibrating Surface Actuators for Active Flow Control*; Proc. of SPIE Vol. 4698, pp. 85–96

[36] Moses R.W., 1997: *Vertical Tail Buffeting Alleviation Using Piezoelectric Actuators – Some Results of the Actively Controlled Response of Buffet-Affected Tails (ACROBAT)*; NASA Technical Memorandum 110336

[37] Simpson J. and J. Schweiger, 1998: *An Industrial Approach to Piezo Electric Damping of Large Fighter Aircraft Components*; 5th Annual Int. Symp. on Smart Structures & Materials, San Diego, CA

[38] Becker J. and L. Luber, 1998: *Comparison of Piezoelectric Systems and Aerodynamic Systems for Aircraft Vibration Alleviation*; 5th Annual Int. Symp. on Smart Structures & Materials, San Diego, CA

[39] Becker J., 2002: *Active Buffeting Vibration Alleviation – Demonstration of Intelligent Aircraft Structure for Vibration and Dynamic Load Alleviation*; Proc. of the ESF-NSF Workshop on Sensor Technology and Intelligent Structures

[40] Simpson J. and C. Boller, 2002: *Performance of SMA-reinforced composites in an aerodynamic profile*; Proc. of SPIE Vol. 4698, pp. 416–426

[41] Barrett R., 1990: *Intelligent rotor blade structures development using directionally attached piezoelectric crystals*; MSc thesis, University of Maryland, College Park, MD

[42] Jänker P., F. Hermle, T. Lorkowski, S. Storm, M. Wettemann and M. Gerle, 2000: *Actuator technology based on smart materials for adaptive systems in aerospace*; Proc. ICAS 2000, Harrogate, UK

[43] Kube R. and V. Klöppel, 2001: *On the role of prediction tools for adaptive rotor system developments*; Smart Mater. Struct., **10**, pp. 137–144

[44] Bebesel M., R. Maier and F. Hoffmann, 2001: *Reduction of Interior Noise in Helicopters by using Active Gearbox Struts – Flight Test Results*; Proc. of 27th Eur. Rotorcraft Forum, Moscow, Russia

[45] Bronowicki A.J., N.S. Abhyanka and S.F. Griffin, 1999: *Active vibration control of large optical space structures*; Smart Mater. Struct., **8**, pp. 740–752

[46] O'Regan S.D., B. Burkewitz, C.R. Fuller, S. Lane and M. Johnson, 2002: *Payload noise suppression using distributed active vibration absorbers*; Proc. of SPIE Vol. 4698, pp. 150–159

[47] Lockyer A.J. *et al.*, 1995: *Development of a Conformal Load Carrying Smart Skin Antenna for Military Aircraft*; Proc. of SPIE Vol. 2448, p. 53

[48] Kiely E., G. Washington and J. Bernhard, 1998: *Design and development of smart microstrip patch antennas*; Smart Mater. Struct., **7**, pp. 792–800

[49] Tennant A. and B. Chambers, 2004: *Adaptive radar absorbing structure with PIN diode controlled active frequency selective surface*; Smart Mater. Struct., **13**, pp. 122–125

[50] Hunt S., A. Rudge, M. Carey, M. Parfitt, J.G. Chase and I. Huntsman, 2002: *Micro-electro-mechanical-systems direct fluid shear stress sensor arrays for flow control*; Smart Mater. Struct., **11**, pp. 617–621

[51] Vinoy K.J. and V.K. Varadan, 2001: *Design of reconfigurable fractal antennas and RF-MEMS for space-based systems*; Smart Mater. Struct., **10**, pp. 1211–1223

[52] Barrett R., 2003: *Developmental History of a New Family of Subscale, Convertible, High Performance UAVs*; Presentation given at the MAV Workshop held at Schloss Elmau

[53] Barrett R.M., C. Burger, J.P. Melian and K. Fidler, 2001: *Recent advances in uninhabited aerial vehicle (UAV) flight control with adaptive aerostructures*; Proc. of the Eur. Conf. on Smart Technology Demonstrators and Devices, Edinburgh, Scotland

[54] Kornbluh R. *et al.*, 2004: *Application of Dielectric Elastomer EAP Actuators*; in Bar-Cohen J. (Ed.) *Electroactive Polymer (EAP) Actuators as Artificial Muscles*; SPIE Press, pp. 529–581

[55] Ho S., H. Nassef, N. Pornsin-Sirirak, Y.C. Tai and C.M. Ho, 2003: *Unsteady Aerodynamics and Flow Control for Flapping Wing Flyers*; Prog. Aerosp. Sci., **39**, pp. 635–681

[56] Seidenschwarz W., Esser J., Niemand S. and Rauch M., 1997: *Target Costing: Auf dem Weg zum marktorientierten Unternehmen*; in Franz K.-P. and Kajüter P. (Eds): *Kostenmanagement – Wettbewerbsvortecile durch, systematische Kostensteuerung*; Schaeffer-Poeschel Verlag, pp. 101–126 (in German)

7
A Summary of Several Studies with Unsymmetric Laminates

Michael W. Hyer[1], Marie-Laure Dano[2], Marc R. Schultz[3], Sontipee Aimmanee[4] and Adel B. Jilani[5]

[1] Department of Engineering Science and Mechanics, Virginia Polytechnic Institute and State University, Blacksburg, VA 24061, USA
[2] Département de Genie Mécanique, Université LAVAL, Québec, Canada, G1K 7P4
[3] Composite Technology Development, Inc., Lafayette, CO 80026, USA
[4] Mechanical Engineering Department, King Mongkut's University of Technology, Thonburi Thoongkru Bangkok 10140, Thailand
[5] Hewlett-Packard Company, 1000 Northeast Circle Blvd, Corvallis, OR 97330, USA

7.1 INTRODUCTION AND BACKGROUND

The purpose of this chapter is to provide an overview of some of the past work focused on understanding the deformation behavior of unsymmetrically laminated plates. By 'unsymmetric' is meant that the thermoelastic properties

Adaptive Structures: Engineering Applications   Edited by D. Wagg, I. Bond, P. Weaver and M. Friswell
© 2007 John Wiley & Sons, Ltd

of the plate are not symmetric with respect to the geometric midplane of the laminate. As a result, rather interesting, and somewhat unusual, elastic couplings occur between inplane and out-of-plane deformations. In particular, the discussions will consider laminated plates that are unsymmetrically laminated and require elevated temperature processing. The elevated temperature may be necessary to cure the material the plate is fabricated from, such as a polymer-matrix fiber-reinforced composite material, or to cure the adhesive bonding the layers together, as will be the case when small piezoceramic-based actuators are discussed. The deformations of interest are those that are thermally induced due to cooling the plates from the elevated curing temperature to the use, or service, temperature. However, when piezoceramic layers are involved, there is interest in the deformations due to actuating the piezoceramic materials. As will be seen, in many of the situations discussed, the thermally induced effects coupled with geometric nonlinearities can result in more than one cooled shape for the plate. The cooled configuration of the plate can be changed from one shape to another by the application of forces or moments. The change from one shape to another is generally a dynamic one and is a snap-through event. At the structural level, this characteristic provides the potential for morphing of the structural configuration. For the small actuators, knowledge of the multiple-shape characteristic is beneficial for tailoring the design either to avoid the pitfalls of multiple shapes, or to take advantage of them.

This chapter begins by briefly reviewing the energy-based approach used to predict the fact that a $[0_2/90_2]_T$ cross-ply graphite–epoxy laminate, processed at an elevated temperature to cure the epoxy, very frequently cools to have two room-temperature shapes, both cylindrical in nature. A laminate such as this, often referred to as an antisymmetric cross-ply laminate, can be snapped from one cylindrical configuration to the other by applying moments along opposite edges of the laminate. In some situations, the laminate cools to have a saddle shape, which is the shape predicted by classical lamination theory as presented in leading textbooks dealing with the mechanics of composite materials. The saddle shape is unique and cannot be snapped through to another shape. The key to whether a single saddle or two cylindrical shapes occur is whether or not geometric nonlinearities are important in the thermal deformation process. When the thermally induced out-of-plane deflections of the originally flat laminated plate are many times the plate thickness, geometric nonlinearities become important. The energy-based approach used to explain the multiple-shape character of an antisymmetric $[0_2/90_2]_T$ cross-ply laminate served as the genesis for a number of later studies, so this particular case is presented first to serve as a tutorial, and to provide a historical perspective.

After $[0_2/90_2]_T$ cross-ply laminates are considered, the discussion turns to the multiple-shape characteristics of more general laminates, such as a $[-30_4/+30_4]_T$ laminate. As will be seen, the concepts and principles employed for determining the cooled shape of these laminates were the same as used to study the $[0_2/90_2]_T$ cross-ply case. Comparisons with empirical results for general laminates are presented.

Next, the discussion of the multiple shapes of unsymmetrically laminated composite plates is connected with current-day interests in the morphing of structures by considering the use of actuators to transform the shape of an unsymmetrically laminated plate from one configuration to another. This concept has merit because energy is expended only to change the shape of a structure rather than use energy continuously to elastically deform a structure to achieve a shape other than its equilibrium configuration. Specifically discussed are the use of shape memory alloy, in the form of wires, and piezoceramic materials, in the form of thin layers, to produce the forces or moments required to effect the snap-through of an unsymmetrically laminated composite plate from one shape to the other.

Finally, the cooled shape and actuated deformations of small piezoceramic actuators are considered. The actuators of interest are manufactured from isotropic metallic layers, or from layers of various composite materials. In either case, the piezoceramic layer, which is essentially isotropic, is sandwiched between these layers. The actuators are manufactured by bonding the layers together with a polymeric adhesive, or by co-curing composite and piezoceramic layers. In both cases elevated-temperature curing and cooling are involved, and in both cases multiple cooled shapes, also cylindrical, can occur. The manufacturers of these small actuators claim that the pre-stressing which results from the elevated-temperature cure, and which produces the cooled shapes, results in better actuation performance than if the actuator was flat. This issue, among others, will be examined.

The chapter ends by summarizing the results that transcend the various applications of unsymmetrically laminated materials that are discussed. As stated, this chapter represents an overview, and thus the reader is referred to referenced publications for details.

7.2   ROOM-TEMPERATURE SHAPES OF SQUARE $(0_2/90_2)_T$ CROSS-PLY LAMINATES

The fact that antisymmetric fiber-reinforced cross-ply laminates, and also other types of unsymmetric laminates, often cool from their cure temperature to have two cylindrical shapes rather than one saddle shape was regarded as

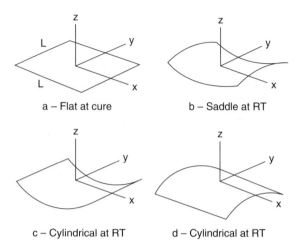

a – Flat at cure   b – Saddle at RT

c – Cylindrical at RT   d – Cylindrical at RT

Figure 7.1 Shapes of an antisymmetric $[0_2/90_2]_T$ cross-ply composite laminate (RT denotes room temperature)

nothing more than a curiosity in the late 1970s and early 1980s. Designers, and even researchers, did not want to deal with the non-flat character of unsymmetric laminates. Also, the fact that curvatures will develop when inplane forces are applied to an unsymmetric laminate was not viewed as a favorable characteristic. The multiple-shape characteristics of unsymmetrically laminated composite plates were unexplained until Hyer (1981a, 1981b, 1982) developed an energy-based approach and showed that the characteristics are due to geometrically nonlinear effects. By considering all the assumptions of classical lamination theory to be valid, including the Kirchhoff hypothesis, but including nonlinear terms in the strain–displacement relations, a relatively simply theory was developed that offered an explanation. The specific problem discussed by Hyer (1981b, 1982) is illustrated in Figure 7.1, where a square laminated plate of sidelength L, flat at its curing temperature and deformed at room-temperature (RT), is illustrated, as is the analysis coordinate system. By the Kirchhoff hypothesis, the inplane strains vary linearly through the thickness of the laminate as

$$\varepsilon_x = \varepsilon_x^0 + z\kappa_x^0; \ \varepsilon_y = \varepsilon_y^0 + z\kappa_y^0; \ \gamma_{xy} = \gamma_{xy}^0 + z\kappa_{xy}^0 \tag{7.1}$$

The midplane strains and curvatures were defined by Hyer (1981b, 1982) as

$$\varepsilon_x^0 = \frac{\partial u^0}{\partial x} + \frac{1}{2}\left(\frac{\partial w^0}{\partial x}\right)^2; \ \varepsilon_y^0 = \frac{\partial v^0}{\partial y} + \frac{1}{2}\left(\frac{\partial w^0}{\partial y}\right)^2; \ \gamma_{xy}^0 = \left(\frac{\partial u^0}{\partial y} + \frac{\partial v^0}{\partial x}\right) + \left(\frac{\partial w^0}{\partial x}\right)\left(\frac{\partial w^0}{\partial y}\right)$$

$$\tag{7.2}$$

$$\kappa_x^0 = -\frac{\partial^2 w^0}{\partial x^2}; \ \kappa_y^0 = -\frac{\partial^2 w^0}{\partial y^2}; \ \kappa_{xy}^0 = -2\frac{\partial^2 w^0}{\partial x \partial y} \tag{7.3}$$

where u^0, v^0, and w^0 are the displacements of the laminate reference surface in the x-, y-, and z-directions, respectively. The underlined terms in Equation (7.2) are the von Kármán approximations to the full nonlinear strain–displacement terms. These are not included in the discussion of classical lamination theory in basic introductory texts (Hyer, 1998).

Hyer (1981b, 1982) used the fact that the shape, or shapes, of the cooled laminate are the shapes that minimize the total potential energy of the laminate. The total potential energy of a cooled laminate is given by

$$\Pi = \frac{1}{2} \int_{-\frac{L}{2}}^{\frac{L}{2}} \int_{-\frac{L}{2}}^{\frac{L}{2}} \int_{-\frac{H}{2}}^{\frac{H}{2}} [(\sigma_x - \sigma_x^T)\varepsilon_x + (\sigma_y - \sigma_y^T)\varepsilon_y + (\sigma_{xy} - \sigma_{xy}^T)\gamma_{xy}]dx\,dy\,dz \tag{7.4}$$

where H is the thickness of the laminate. The $\overline{Q}$s are the transformed reduced stiffnesses which relate the stresses and strains as

$$\begin{aligned}
\sigma_x &= \overline{Q}_{11}\varepsilon_x + \overline{Q}_{12}\varepsilon_y + \overline{Q}_{16}\gamma_{xy} - \sigma_x^T \\
\sigma_y &= \overline{Q}_{12}\varepsilon_x + \overline{Q}_{22}\varepsilon_y + \overline{Q}_{26}\gamma_{xy} - \sigma_y^T \\
\sigma_{xy} &= \overline{Q}_{16}\varepsilon_x + \overline{Q}_{26}\varepsilon_y + \overline{Q}_{66}\gamma_{xy} - \sigma_{xy}^T
\end{aligned} \tag{7.5}$$

The σ^Ts are the thermally induced stresses given by

$$\begin{aligned}
\sigma_x^T &= \overline{Q}_{11}\varepsilon_x^T + \overline{Q}_{12}\varepsilon_y^T + \overline{Q}_{16}\gamma_{xy}^T \\
\sigma_y^T &= \overline{Q}_{12}\varepsilon_x^T + \overline{Q}_{22}\varepsilon_y^T + \overline{Q}_{26}\gamma_{xy}^T \\
\sigma_{xy}^T &= \overline{Q}_{16}\varepsilon_x^T + \overline{Q}_{26}\varepsilon_y^T + \overline{Q}_{66}\gamma_{xy}^T
\end{aligned} \tag{7.6}$$

The thermally induced dilatational strains, or free thermal strains, are

$$\begin{aligned}
\varepsilon_x^T &= \alpha_x \Delta T \\
\varepsilon_y^T &= \alpha_y \Delta T \\
\gamma_{xy}^T &= \alpha_{xy} \Delta T
\end{aligned} \tag{7.7}$$

The temperature change from the cure temperature to room temperature is ΔT. The notation used is standard for classical lamination theory.

By integrating Equation (7.4) with respect to z, the well-known A, B, and D matrices are defined. By using variational techniques, it is possible to derive the nonlinear differential equations and boundary conditions, in terms of the three components of displacement, that govern the deformation behavior of the laminate as it is cooled from its curing temperature. However, solutions to the differential equations and boundary conditions would be difficult to obtain, so approximate solutions in the sense of Rayleigh–Ritz were sought. With that approach, the functional forms for the three displacement components were assumed, and the total potential energy minimized within the context of those approximate deformations. In that regard, the out-of-plane deflections, $w^0(x, y)$, of the three laminated plates in Figures 7.1(b–d) can be represented in an approximate sense by the functional form

$$w^0(x, y) = \frac{1}{2}(ax^2 + by^2) \tag{7.8}$$

where a and b are unknown but to-be-determined coefficients related to the curvatures of the deformed plate (see Equation (7.3)). With $b = -a$ the saddle of Figure 7.1(b) is represented, with $a > 0$, $b = 0$ the cylindrical shape of Figure 7.1(c) is approximated, and with $a = 0$, $b < 0$ the cylindrical shape of Figure 7.1(d) is approximated. Following the logic in Hyer (1982), the other two components of displacement are approximated by

$$
\begin{aligned}
u^0(x, y) &= cx - \frac{a^2x^3}{6} - \frac{abxy^2}{4} \\
v^0(x, y) &= dy - \frac{b^2y^3}{6} - \frac{abx^2y}{4}
\end{aligned}
\tag{7.9}
$$

where c and d are two other to-be-determined coefficients. Equations (7.8) and (7.9) are substituted in Equations (7.2), (7.3), and (7.1), then (7.5) and finally (7.4), and the integrations with respect to x and y carried out explicitly. The expression for total potential energy is then an algebraic expression in terms of laminate material properties, i.e., the components of the A, B, and D matrices, and geometric properties, and, of course, the coefficients a, b, c, and d. As total potential energy is a minimum at equilibrium, the first variation of the total potential energy with respect to variations in the coefficients a, b, c, and d is computed. The values of a, b, c, and d that satisfy the resulting four conditions, which are nonlinear algebraic equations, are then determined. Since the equations are nonlinear, multiple solutions, corresponding to multiple cooled shapes, could be expected. Since the values of a, b, c, and d that satisfy the resulting four conditions could also satisfy energy maximums, and thus unstable shapes, the second variation of the total potential energy is examined for positive definiteness as a stability check.

Considering a medium-modulus graphite–epoxy material and allowing the sidelength L of the square laminate in Figure 7.1(a) to vary, the relationships between the curvature parameters a and b at room temperature and the sidelength L illustrated in Figure 7.2 result. As seen, for sidelengths less

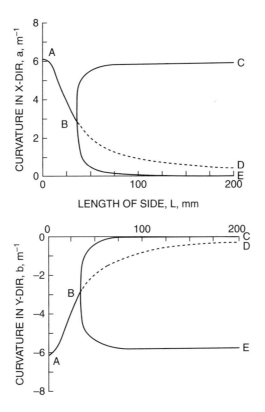

Figure 7.2 Variation of curvature parameters a and b with sidelength L of square antisymmetric $[0_2/90_2]_T$ cross-ply graphite–epoxy laminate: solid, stable; dashed, unstable

than about 37.5 mm, the relationships are single valued, while for sidelengths greater 37.5 mm the relationships are multi-valued. The relationships bifurcate at $L = 37.5$ mm, point B in the figure. This implies that for larger laminates there is more than one cooled shape. The relationship represented by branch BD represents unstable cooled shapes, and thus will not be observed. On the other hand, stable branch BC represents a cooled laminate with curvature in the x-direction and no curvature in the y-direction, a cylindrical shape. Stable branch BE represents a cooled laminate with curvature in the y-direction and no curvature in the x-direction, another cylindrical shape. The curvatures of the shapes represented by branches BC and BE have the same magnitude but opposite sign. Branch AB represents saddle shapes, with equal and opposite curvatures. The curvatures at $L = 0$ are the predictions obtained without including the nonlinear terms in the strain–displacement relations, Equation (7.2), and are thus the predictions of classical lamination theory. It is seen that for sidelengths greater than zero, the curvatures of the saddle are suppressed relative to the linear prediction and the cooled saddle-shaped laminate becomes flatter and flatter as the sidelength increases until the laminate becomes unstable.

It should be mentioned that fabricating antisymmetric laminates can be difficult. Constructing laminates to have fiber orientations exactly 0 degrees or exactly 90 degrees is almost impossible. Also, it is difficult to have all layers exactly the same thickness, as thickness differences result from slight variations in the composite pre-preg as well as from resin bleed during curing. However, these factors do not preclude the existence of multiple shapes. A discussion of the role of imperfections, and a slightly different view of antisymmetric cross-ply laminates, is available in Hamamoto and Hyer (1987). Also, one of the first studies of the cooled shapes of unsymmetric laminates based on finite-element analysis is discussed by Schlecht *et al.* (1995).

7.3 ROOM-TEMPERATURE SHAPES OF MORE GENERAL UNSYMMETRIC LAMINATES

The cooled shapes of more general unsymmetric laminates were studied by generalizing the approach taken to study antisymmetric cross-ply laminates. Dano and Hyer (1998) expanded on the above development for cross-ply laminates by assuming a functional form of the inplane extensional strains and accounting for the possibility of twist curvature in the expression for the out-of-plane displacement. Specifically, the forms assumed were

$$\varepsilon_x^0(x, y) = c_1 + c_2 x^2 + c_3 y^2 + c_4 xy$$
$$\varepsilon_y^0(x, y) = c_5 + c_6 x^2 + c_7 y^2 + c_8 xy \qquad (7.10)$$
$$w^0(x, y) = \tfrac{1}{2}\left(c_9 x^2 + c_{10} y^2 + c_{11} xy\right)$$

The coefficients c_1, c_2, etc., are to-be-determined coefficients. The functional form for the inplane shear strain $\gamma_{xy}^0(x, y)$ is determined by integrating the strain expressions of Equation (7.10) to find expressions for the inplane displacements u^0 and v^0, insuring displacement compatibility, and using the third expression in Equation (7.2). The net result is the introduction of 3 more unknown coefficients, for a total of 14 unknown coefficients. The assumed functional forms are substituted into Equations (7.3), (7.1), (7.5), and finally (7.4), which, as written, includes general laminates, and integration with respect to x and y is carried out. The 14 coefficients representing stable cooled shapes are determined by minimizing the total potential energy with respect to these coefficients by way of the first variation. As with the cross-ply laminates, solutions to these 14 nonlinear algebraic equations also give cooled shapes that corresponded to an energy maximum, and thus unstable conditions. Stability of the cooled shape is evaluated by checking for positive definiteness of the second variation of the total potential energy for solution values of the 14 coefficients.

As a check on the Rayleigh–Ritz solution approach, a finite-element analysis of the thermally induced deformations was developed using the commercial code ABAQUS and a 10 by 10 mesh of four-node shell elements. Finite-element approaches often have difficulty finding multiple solutions to nonlinear problems, and that was the case for this problem. Here it was sufficient to compare just one of the cooled shapes as predicted by the finite-element analysis with the shape predicted by the Rayleigh–Ritz solution.

Several general 0.254 m by 0.254 m (10 in by 10 in) laminates were fabricated using graphite–epoxy and the out-of-plane deflections of the cooled shapes measured. To aid in the measurement and overall observations of the cooled shapes, single light-colored Kevlar™ fibers were embedded in grid fashion on the two surfaces of each laminate. The out-of-plane deflections of the laminates at each grid point were measured to compare with predictions. Qualitative comparisons of the observed shapes and the shapes predicted by the Rayleigh–Ritz solution for $[-30_4/+30_4]_T$ and $[60_4/30_4]_T$ laminates are illustrated in Figures 7.3 and 7.4, respectively. The grid of Kevlar fibers on each of the manufactured laminates is seen. Both stable shapes are shown for both laminates, and it is seen that the qualitative comparison is very good. Quantitative comparisons among the Rayleigh–Ritz solution, the finite-element analysis, and the measured deformations for one of the shapes of each of these two laminates are shown in Figures 7.5 and 7.6. In both cases, the Rayleigh–Ritz solution and the finite-element analysis are in good agreement, despite the finite-element analysis having many more degrees of freedom to describe the deformed laminate than the Rayleigh–Ritz solution. In Figures 7.5 and 7.6 the grids for the Rayleigh–Ritz solution, the finite-element predictions, and the measured shapes become ill-defined in certain regions because the three-dimensional renderings of each of the three surfaces partially block each other, obscuring the grid lines. Note also the reversing of the sense of the x- and y-coordinates in Figures 7.5 and 7.6 relative to Figures 7.3 and 7.4, respectively.

It can be concluded from Figures 7.3–7.6 that the existence of multiple cooled shapes of unsymmetric laminates can be modeled, and the deflections associated with these shapes can be predicted with reasonable accuracy. The next section considers changing the shape of unsymmetric laminates from

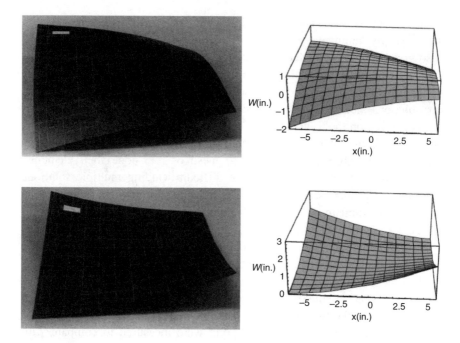

Figure 7.3 Observed (left) and predicted (right) cooled shapes of 0.254 m by 0.254 m (10 in by 10 in) $[-30_4/+30_4]_T$ graphite–epoxy laminate

one configuration to the other by way of applied forces in the form of moments.

7.4 MOMENTS REQUIRED TO CHANGE SHAPES OF UNSYMMETRIC LAMINATES

It is the logical next step to ask what force or moment levels are required to produce the snap-through. The multiple stable shapes of unsymmetric laminates have the potential on a larger scale for application to the morphing of structures, and energy need only be expended to change the shape of the laminates, not to hold them in one configuration or the other, as some morphing concepts require. To answer the questions related to the concept, investigations with small-scale laminates continued and the means to snap a laminate from one configuration to the other by way of an actuator were studied.

On the scale of the laminates considered above, logical actuation schemes to consider were those based on shape memory alloys or piezoceramic

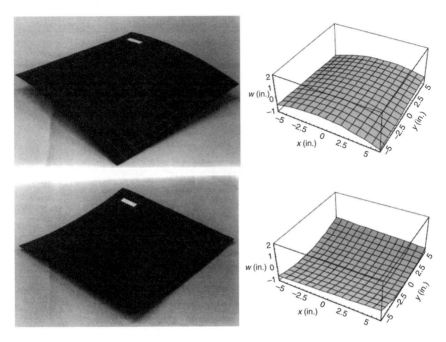

Figure 7.4 Observed (left) and predicted (right) cooled shapes of 0.254 m by 0.254 m (10 in by 10 in) $[60_4/30_4]_T$ graphite–epoxy laminate

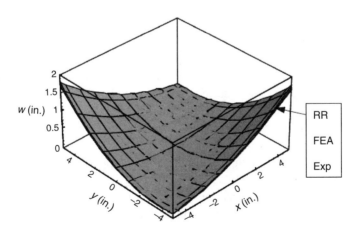

Figure 7.5 Comparison between the Rayleigh–Ritz solution (RR), finite-element analysis (FEA), and experimental measurements (Exp) for 0.254 m by 0.254 m (10 in by 10 in) $[-30_4/30_4]_T$ graphite–epoxy laminate

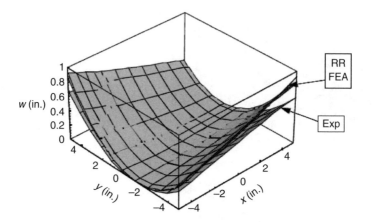

Figure 7.6   Comparison between the Rayleigh–Ritz solution (RR), finite-element analysis (FEA), and experimental measurements (Exp) for 0.254 m by 0.254 m (10 in by 10 in) $[60_4/30_4]_T$ graphite–epoxy laminate

materials. There were advantages or disadvantages to each type of material. For shape memory alloys the issues were: Did they have the authority to effect the shape change? Should they be on the surface of the laminate or embedded within? Was the thermoelastic behavior of shape memory alloys understood well enough to utilize them? Was the behavior repeatable? For piezoceramic materials the issues were similar: Did they have the authority to effect the shape change? Should they be on the surface of the laminate or embedded within? Were they too brittle for the curvatures involved? Piezoceramic materials generally are associated with dynamic applications. Would they work for the quasi-static application here? Both approaches were considered and will be discussed (Dano and Hyer, 2003; Schultz and Hyer, 2003). First, however, it was deemed necessary to know what moment levels were required to change an unsymmetric laminate from one shape to another.

A simple experiment conducted to determine the force levels required to snap the laminate from one shape to the other is illustrated in Figure 7.7 (Dano and Hyer, 2002). In this experiment, three supports, labeled supports a, b, and c, were fastened to the laminate, a $[0_4/90_4]_T$ laminate being illustrated in Figure 7.7. Support a was considerably longer than the others for the purposes of attaching the laminate–support arrangement to a stiff base and suspending the arrangement as shown. A thin steel wire was fixed to support c a known distance from the surface of the laminate. The wire wrapped around support b at the same known distance from the surface of the laminate and could slide around the support as it stretched and the laminate deformed. The wire then wrapped and slid around support c and was ultimately attached

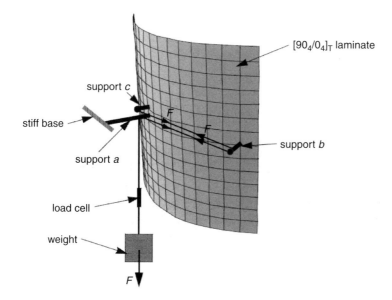

Figure 7.7 Schematic of experiment to measure the force level to change the shape of an unsymmetric laminate

to a known force in the form of a known weight. Grooves in the supports guaranteed that the wire maintained a fixed distance from the surface of the laminate. With this arrangement, the tension in the wire between supports b and c, combined with the distance away from the surface of the laminate, exerted a known moment on the laminate. When the moment was of sufficient magnitude, the laminate would snap from one shape to the other. With this scheme, the magnitude of the moment required to snap the laminate could be accurately determined if the weight and distance of the wire from surface of the laminate were accurately known.

In the experiment, the force, or weight, was actually due to water slowly draining into a container and causing the force to increase in a controlled fashion. This approach resulted in a force-controlled experiment, as opposed to a displacement-controlled experiment. The force to cause the laminate to snap was determined by measuring the weight of the water. As a check, a load cell in series with the steel wire also monitored the force in the wire.

To develop an analysis to predict the force level to produce snapping, the principle of virtual work, in conjunction with the Rayleigh–Ritz technique, was used. Formally, the principle of virtual work can be written as

$$\delta W_T = \delta \Pi - \delta W_F \tag{7.11}$$

where δW_T is the total virtual work, $\delta\Pi$ the internal virtual work, and δW_F the virtual work of the applied force. Of course $\delta\Pi$ is the first variation of the total potential energy of Equation (7.4). The virtual work of the applied force is given by

$$\delta W_F = \vec{F} \cdot \delta R_F\big|_b + \vec{F} \cdot \delta R_F\big|_c \qquad (7.12)$$

where δR_F is the virtual displacement of the support at the point of application of the wire force on the support, and subscripts b and c identify the support. The quantity $\vec{R}_F$ is the position vector to the point of application of the force. Important geometric quantities needed for the involved calculation of the virtual displacement δR_F are illustrated in Figure 7.8 for a representative support. Referring to the figure, the position vector $\vec{R}_F$ is given by

$$\vec{R}_F = \vec{r} + \vec{n}^* \qquad (7.13)$$

In this equation, $\vec{r}$ is the position vector to the base of the support and $\vec{n}^*$ is the position vector from the base of the support to the point of application of

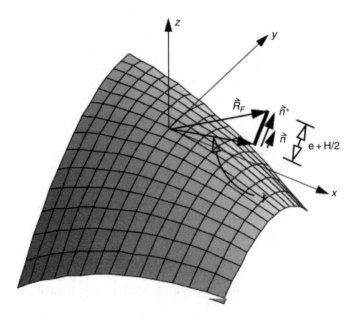

Figure 7.8 Geometric quantities needed to compute virtual displacement

the force on the support. The support is assumed to be rigid, so $\vec{n}^*$ is given by

$$\vec{n}^* = \left(e + \frac{H}{2} \right) \vec{n} \tag{7.14}$$

where e is the distance along the support from the laminate surface to the groove in the support, namely the point of application of the force, and $H/2$ is the half the laminate thickness, the sum representing the distance from the laminate reference surface to the point of application of the force. The quantity $\vec{n}$ is the unit normal to the reference surface at the support location. It is given by

$$\vec{n} = \frac{\dfrac{\partial \vec{r}}{\partial x} \times \dfrac{\partial \vec{r}}{\partial y}}{\left| \dfrac{\partial \vec{r}}{\partial x} \times \dfrac{\partial \vec{r}}{\partial y} \right|} \tag{7.15}$$

where the vector $\vec{r}$ can be written as

$$\vec{r} = \left(x + u^0(x, y) \right) \hat{i} + \left(y + v^0(x, y) \right) \hat{j} + w^0(x, y) \hat{k} \tag{7.16}$$

In Equation (7.16) $\hat{i}$, $\hat{j}$, and $\hat{k}$ are unit vectors in the x-, y-, and z-directions, respectively. As seen, the expression for $\vec{r}$, and thus $\vec{n}$ and in turn $\vec{n}^*$, involves the three displacement components. The displacement components are assumed to have the same functional forms as discussed in relation to Equation (7.10) for a general laminate and involve 14 unknown coefficients. Accordingly, the expression for $\vec{n}$ is quite complicated. However, the net result is a prediction of the deformed shape of the laminate due to cooling plus the application of the known force. The shape and stability characteristics of the laminate as a function of the force, or moment, level can be computed from the developed analysis and the force level required for snap-through predicted.

In the experimental set-up shown schematically in Figure 7.7, back-to-back strain gages were mounted at the geometric center of each laminate to monitor the force level at which snap-through occurred. As the force was applied, the strain gages recorded moderate changes in strain levels. A large change in strain levels signaled snap-through. In Figure 7.9 the strain vs. moment history for a $[90_4/0_4]_T$ laminate is illustrated, the force level being converted to a moment level using the geometry of the arrangement of Figure 7.7. The predicted and measured strain vs. moment relations are shown. The experiments were done several times to evaluation repeatability.

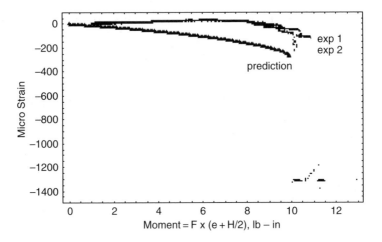

Figure 7.9 Moment vs. strain level for an antisymmetric $[90_4/0_4]_T$ cross-ply graphite–epoxy laminate ($1 \, \text{lb-in} = 0.1130 \, \text{N} \cdot \text{m}$)

As seen in Figure 7.9, as the moment level was increased from zero in the experiment, the strain level remained relatively constant. At a moment level in the range 1.186 to 1.243 N·m (10.5 to 11.0 lb-in), the laminate snapped. The predicted level was close to 1.130 N·m (10 lb-in). The analysis predicted a gradual change in strain to the 1.130 N·m (10 lb-in) moment level, then a sudden change. The strains measured during the experiments did not change as much when the moment level was increased. One factor contributing to the lack of change in the measured strains relative to the prediction was that the analysis assumed all locations of the laminate had the same curvature (see the expression for $w^0(x, y)$ in Eq. (7.10)). In a sense, this single curvature value could be thought of as the average laminate curvature. In reality, the supports transmitted the moment to the laminate where the supports were fastened. There was no doubt local bending, and thus increased strain levels, at those locations not registered by the strain gages in the center of the laminate. However, the correlation between measurements and predictions was quite reasonable, and it was believed an understanding of the moment levels required was in hand. A variant of this experiment is described in Dano and Hyer (1996).

7.5 USE OF SHAPE MEMORY ALLOY FOR ACTUATION

The experiments and developed analysis just discussed were necessary to implement the concept of using shape memory alloy (SMA) to force the laminate from one shape to the other. SMA is available in a variety of forms. The general characteristics of the stress vs. temperature relation of the material are known. Having an estimate of the moment level needed to change the shape provided guidance as to the form of SMA to use, how to integrate it with the laminates, and other considerations. For the work here, it was decided to use SMA in wire form, as it is easy to work with and is readily available. From the force levels discussed in the previous section, it did not seem possible to generate the moment levels necessary if the SMA wires were embedded within the laminate. Additionally, the issue of attaining a strong and reliable bond between the SMA wire and the graphite–epoxy material was in question. Also, it would be difficult to change the diameter of the wire, for example, or to calibrate the wire once it was embedded within the laminate. An example of how the SMA wire was used is shown in Figure 7.10. In the figure it is seen that five supports were mounted on a $[-30_4/30_4]_T$ graphite–epoxy laminate, much as was done for the experiments described in the previous section and depicted schematically in Figure 7.7. A length of thin SMA wire was fixed at the two outer supports of the three-support group near one corner of the laminate and passed around the two supports near the opposite corner, and around the middle support of

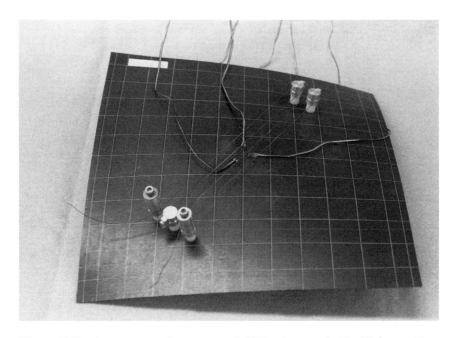

Figure 7.10   Arrangement of supports and SMA wire on a $[-30_4/30_4]_T$ graphite–epoxy laminate

the three-support group. Again, grooves in the supports guaranteed that the wire remained at a fixed distance from the surface of the laminate. With a single wire fixed at both ends and passing around each support, and freely sliding, the tension along the entire length of wire was the same, and the more times the wire passed back and forth between supports at opposite locations on the laminate, the more moment could be produced with a single wire. The direction of the wires relative to the edges of the laminate were a function of the lamination sequence, and were oriented to be aligned with one of the principal curvature directions. As the directions and magnitudes of the principal curvatures, and thus the magnitudes of moments required to snap the laminates, varied with lamination sequence, the number of supports and their location varied from laminate to laminate, as illustrated in Figure 7.11 for the four laminates studied.

Several other parameters had to be considered when using SMA wires with the approach described. To use SMA, it is generally pre-strained and then heated. The heating produces the required change of phase and associated increase in stress at that strain level, in this case a tensile stress, and a resulting force, in this case on the supports the wire was fixed to and wrapped around.

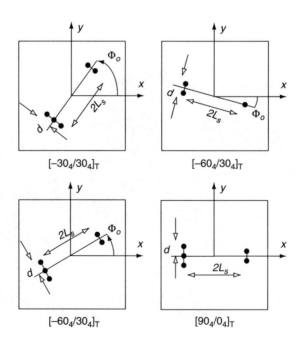

Figure 7.11 Arrangements of supports and SMA wire for various lamination sequences (Φ_o = principal curvature direction)

With the particular wire used, an 8 % pre-strain level was used. Because the supports moved as the curvature of the laminate changed due to the stress in the wire, it was necessary to keep the geometry of the supports and laminate curvature such that the 8 % level strain was not totally recovered before the laminate snapped. Additionally, as it was required that the wires remain straight and stretched between the supports, it was important that they did not touch the curved laminate. Finally, since electric current was used to heat the SMA wires, and since graphite–epoxy can conduct current, it was necessary to electrically insulate the supports from the laminate.

For reference, a schematic of the variation of stress with temperature for SMA is illustrated in Figure 7.12. The well-known martensitic–austenitic phase change as the temperature increases is responsible for the increased stress level, for a fixed strain level, as illustrated. Specifics of the constitutive equation used to represent the SMA wire, and the calibration procedure, are discussed in Dano and Hyer (2003).

As seen in Figure 7.10, strain gages were mounted at the center of the laminate to serve as an indicator of snapping, as was done with the controlled force experiments described earlier. The relation between one of the measured strains and temperature for a $[-60_4/30_4]_T$ laminate is illustrated in Figure 7.13. The predicted strains and strains as measured during repeated tests are shown. Temperature was measured with a thermocouple mounted on the wire. As observed, the correlation between prediction and observation is quite good. The spread between the repeat tests is reasonable,

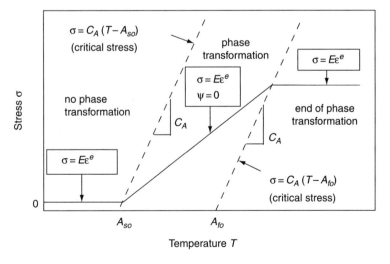

Figure 7.12 Schematic of stress vs. temperature relation for SMA at a fixed strain level

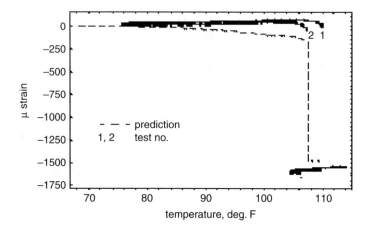

Figure 7.13   Temperature vs. strain relations for $[-60_4/30_4]_T$ graphite–epoxy laminate (100 deg F = 37.8 degrees C)

and like the controlled force experiment, the predicted strains changed more than the measured strains up to the point of snap-through. This is again attributed to assuming that laminate deformations could be described by a single curvature, and local flexiblilty of the laminate where the support was attached.

7.6   USE OF PIEZOCERAMIC ACTUATION

The development of the predictive analysis and implementation of SMA to produce the moments necessary to snap the laminate was a success. Unfortunately, the SMA wire–support arrangement was somewhat clumsy and probably limited the application of the concept. Attention then turned to the use of piezoceramic material bonded to the surface of the laminate, specifically in the form of a Macro-Fiber Composite™ (MFC™) actuator (Wilkie *et al.*, 2000) bonded to the surface. Such an arrangement would be much more streamlined, compact, and aesthetically appealing. As the antisymmetric cross-ply lamination sequence required the most moment to snap the laminate, attention was limited to this laminate.

   The two shapes of a two-layer antisymmetric $[0/90]_T$ cross-ply graphite–epoxy laminate with the MFC actuator bonded on one side are shown in Figure 7.14. The laminate was approximately 0.150 m by 0.150 m (6 in by 6 in) and, as seen, the MFC actuator covered a reasonable portion of the laminate. In terms of forces and moment arms required to snap the laminate, with the actuator bonded to the surface, the moment arm was about half

(a) – One equilibrium shape

(b) – Other equilibrium shape

Figure 7.14 Antisymmetric $[0/90]_T$ cross-ply graphite–epoxy laminate with Macro-Fiber Composite™ (MFC™) actuator (x-axis out of page, y-axis to the right)

the thickness of the laminate. Therefore either the actuator had to be quite powerful, or it needed to cover considerable area. The wires to the MFC electrodes are visible in Figure 7.14, as is the wire to the strain gage mounted in the center of the laminate on the backside.

The MFC actuator is a unique device. Its operation and manufacture will not be discussed here. Its construction and features are discussed by Wilkie *et al.* (2000), with further details given by Williams *et al.* (2004). However, the exploded view of the actuator given in Figure 7.15 provides insight into the device. Basically the actuator is constructed of a standard piezoceramic wafer sliced into narrow pieces with parallel cuts of a thin saw blade. The narrow pieces are referred to as macro-fibers, and they are sandwiched between a set of interdigitated electrodes designed so the electric field is aligned with the long direction of the macro-fibers. This is opposed

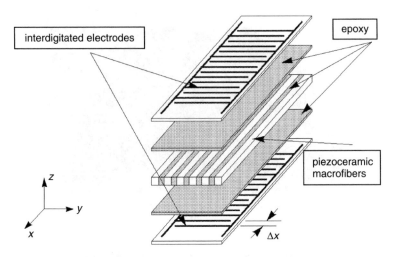

Figure 7.15   Schematic of Macro-Fiber Composite™ (MFC™) actuator

to having the field pass through the thickness direction of the original wafer, as is often the orientation of the electric field when using piezoceramic materials. The MFC actuator has two advantages. First, the MFC actuator is much more flexible than a single piezoceramic wafer, and, second, its actuation characteristics are orthotropic. The flexibility makes it practical for application to curved surfaces, such as an unsymmetric laminate, and the orthotropic properties seem to fit well with orthotropic properties of fiber-reinforced composite materials.

The MFC actuator could be considered just another layer in the laminate, though it was only a partial layer since it did not extend to the edges of the laminate. The mechanical and piezoelectric properties of the actuator could be determined, the former by mechanical testing (Williams *et al.*, 2004) and the latter by calibration (Schultz and Hyer, 2003), and these properties integrated into a laminate analysis, much like the inclusion of any other layer in a laminated composite. The actuation-induced strains could be represented as dilatational strains, just like thermal expansion strain, in the principal material coordinate system as

$$\varepsilon_x^E = d_{11}\frac{\Delta V}{\Delta x} = d_{11}E_1, \ \varepsilon_y^E = d_{12}\frac{\Delta V}{\Delta x} = d_{12}E_1, \ \gamma_{xy}^E = 0 \qquad (7.17)$$

where the superscript E reflects the fact the strains were due to the electric field, ΔV is the voltage applied to the actuator, Δx is the spacing between the interdigitated electrodes, and E_1 the notation for electric field strength. The constants d_{11} and d_{12} are piezoelectric material properties and, as written, assume that the macro-fibers are aligned with the x-axis, as shown in Figure 7.15. There is a voltage limit to all piezoelectric devices, as too high a voltage will de-pole the actuator, and for the MFC, too high a voltage could short-circuit the electrodes.

The MFC actuator was bonded to the curved unsymmetric laminate by placing the laminate and actuator, with adhesive on what would be the mating surfaces, inside a vacuum bag and evacuating the bag to pull the laminate and actuator into contact with each other until the adhesive cured at room temperature. The orientation of the laminate and actuator when they were bonded together, the so-called as-bonded shape, were as illustrated in Figure 7.14(a). Since the MFC actuator was initially flat, the vacuum deformed it to bring it into contact with the curved laminate. Since an MFC actuator has bending stiffness, the curvature of the laminate decreased slightly, the result being that the curvature of the laminate–actuator combination was unknown, and, furthermore, there were multiple shapes to the laminate–actuator combination, just as with the laminate alone. With the MFC actuator bonded to the surface, as the voltage in the actuator was increased, the laminate–actuator combination deformed. The goal was to increase the voltage until the laminate–actuator combination with one shape snapped to the other shape. An important objective of this work, in addition to determining whether or not snapping could be induced, was to determine the voltage levels and limits on the size of laminate one actuator could control. This could be done empirically, but another objective of the work was to develop an analysis tool to be able to predict the behavior of the laminate–actuator combination, including the voltage level required to produce snapping.

Development of the analysis tool required consideration of three steps. Step 1 was the cooling of the laminate from the cure temperature to the multiple cooled shapes. This step in the analysis was certainly not new at this point in time. Step 2 was the bonding of the actuator to the laminate and computing the multiple shapes of the laminate–actuator combination. This step required additional analysis development. Step 3 was the actuation of the MFC by way of applied voltage. This step also required additional analysis development. For step 2, a total potential energy similar in form to Equation (7.4) was appended with the expression for the strain energy in the actuator deformed, relative to its flat natural shape, to conform with the laminate, which, as stated earlier, was deformed slightly relative to the cooled shape.

The geometry of the laminate–actuator combination is illustrated in Figure 7.16, where the sidelengths of the laminate and actuator are L_x and L_y, and L_x^{MFC} and L_y^{MFC}, respectively, and the location of the layer interfaces

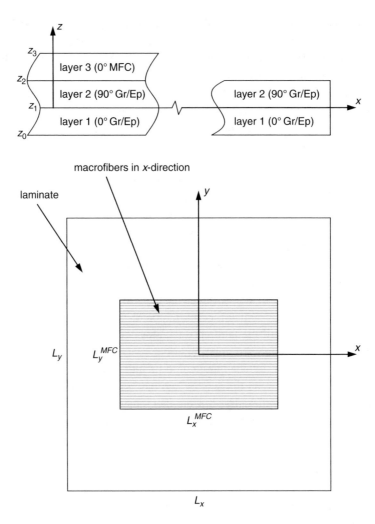

Figure 7.16 Geometry of laminate–actuator combination

in the z-direction are denoted as z_0, z_1, z_2, and z_3. The total potential energy for step 1 was given by

$$\Pi = \Pi_1 = \frac{1}{2} \int_{-\frac{L_x}{2}}^{\frac{L_x}{2}} \int_{-\frac{L_y}{2}}^{\frac{L_y}{2}} \int_{z_0}^{z_2} [(\sigma_x - \sigma_x^T)\varepsilon_x + (\sigma_y - \sigma_y^T)\varepsilon_y + (\sigma_{xy} - \sigma_{xy}^T)\gamma_{xy}] dx\, dy\, dz \qquad (7.18)$$

This equation is similar to Equation (7.4) except the limits for z reflect the fact the actuator was not included for this step. For step 2, the potential energy of the actuator was appended to the potential energy for the laminate, i.e.,

$$\Pi = \Pi_1 + \Pi_2 \tag{7.19}$$

with

$$\Pi_2 = \frac{1}{2} \int_{-\frac{L_x^{MFC}}{2}}^{\frac{L_x^{MFC}}{2}} \int_{-\frac{L_y^{MFC}}{2}}^{\frac{L_y^{MFC}}{2}} \int_{z_2}^{z_3} [\sigma_x^{a2} \varepsilon_x^{a2} + \sigma_y^{a2} \varepsilon_y^{a2} + \sigma_{xy}^{a2} \gamma_{xy}^{a2}] dx\,dy\,dz \tag{7.20}$$

where

$$\begin{aligned}
\sigma_x^{a2} &= \overline{Q}_{11} \varepsilon_x^{a2} + \overline{Q}_{12} \varepsilon_y^{a2} + \overline{Q}_{16} \gamma_{xy}^{a2} \\
\sigma_y^{a2} &= \overline{Q}_{12} \varepsilon_x^{a2} + \overline{Q}_{22} \varepsilon_y^{a2} + \overline{Q}_{26} \gamma_{xy}^{a2} \\
\sigma_{xy}^{a2} &= \overline{Q}_{16} \varepsilon_x^{a2} + \overline{Q}_{26} \varepsilon_y^{a2} + \overline{Q}_{66} \gamma_{xy}^{a2}
\end{aligned} \tag{7.21}$$

and the ε^{a2}s and γ_{xy}^{a2}s denote the strains in the actuator. As seen, the limits on z reflect just the thickness of the actuator. (See Figure 7.16.) For step 3, the total potential energy was

$$\Pi = \Pi_1 + \Pi_3 \tag{7.22}$$

with

$$\begin{aligned}
\Pi_3 = \frac{1}{2} \int_{-\frac{L_x^{MFC}}{2}}^{\frac{L_x^{MFC}}{2}} \int_{-\frac{L_y^{MFC}}{2}}^{\frac{L_y^{MFC}}{2}} \int_{z_2}^{z_3} [(\sigma_x^{a3} - \sigma_x^{E}) \varepsilon_x^{a3} + (\sigma_y^{a3} - \sigma_y^{E}) \varepsilon_y^{a3} \\
+ (\sigma_{xy}^{a3} - \sigma_{xy}^{E}) \gamma_{xy}^{a3}] dx\,dy\,dz
\end{aligned} \tag{7.23}$$

The stresses and strain in the above integrand account for the stresses in the actuator due to bonding it to the laminate, and the stresses induced by energizing the piezoceramic material, as in Equation (7.17). The expressions were complicated and could be written several ways. Details of and the procedures for using the energy expressions can be found in Schultz and Hyer (2003).

A Rayleigh–Ritz approximate solution approach was again used for all three steps of the analysis. However, quarter symmetry of the problem was invoked in selecting the assumed displacement field. This was necessary to keep the number of nonlinear algebraic equations to be solved as a function of voltage level to a minimum. In that regard, the out-of-plane displacement for all three steps was taken to be of the form

$$w^0(x, y) = \frac{1}{2} (c_1 x^2 + c_2 y^2) + \frac{1}{6} (c_3 x^3 + c_4 y^3) \tag{7.24}$$

With the quarter-symmetry assumption, care had to be taken in evaluating the energy integrals. As can be seen from the form of $w^0(x, y)$ in Equation (7.24), the curvature of the laminate was not assumed to be constant over the entire area of the laminate–actuator combination. This was because of the partial coverage of the laminate by the actuator. For the cooling step, step 1, c_3 and c_4 were not of any significance, but for steps 2 and 3, they contributed to the solution. The assumed forms for $u^0(x, y)$ and $v^0(x, y)$ can be found in Schultz and Hyer (2003).

To determine the voltage predicted by the analysis, the values of the four constants c_1, c_2, c_3, and c_4 were computed as a function of the applied voltage. These relations are illustrated in Figures 7.17 and 7.18 and are interpreted as follows. With no voltage applied to the actuator and the laminate–actuator combination in the configuration of Figure 7.14(b), $c_1 \approx -1 \, \text{m}^{-1}$ and $c_3 \approx +95 \, \text{m}^{-2}$, while c_2 and c_4 are approximately zero. As the voltage increases, the values of c_1 and c_3 change, while the values of c_2 and c_4 remain close to zero. At a voltage of 1262 V, the relations for the four coefficients reach a limit point. With a further increase in voltage the relations must suddenly change, or 'snap', to the other branches represented by c_1 and c_3 being practically zero and $c_2 \approx -5.5 \, \text{m}^{-1}$ and $c_4 \approx -45 \, \text{m}^{-2}$, representing the laminate in the shape shown in Figure 7.14(a). Returning the voltage to zero, the shape of Figure 7.14(a) is retained. The transformation of laminate shape has been made.

Using a voltage-controlled power supply, a data acquisition system, and monitoring the strain gage at the center of the laminate, voltage was applied to the actuator with the shape of Figure 7.14(b). As the voltage was increased, very little deformation was recorded until a voltage level of around 1700 V was reached. At this voltage level, the laminate snapped to the configuration of Figure 7.14(a). The voltage was then returned to zero. This procedure was repeated several times with basically the same results. A comparison of the predicted and observed shapes of the laminate is illustrated in Figure 7.19. The qualitative comparison is good, as is the quantitative comparison. Though the correlation between the predicted and measured voltage to produce snap-through was not as good as desired, the experiment and extended analysis were considered a success.

7.7 CONSIDERATION OF SMALL PIEZOCERAMIC ACTUATORS

'Bender' is the name given to small active devices constructed by bonding one or more layers of a piezoceramic material to one or more layers of,

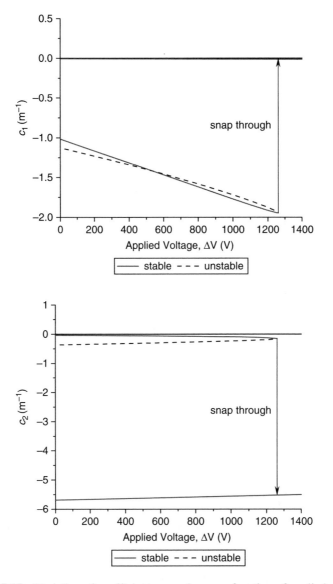

Figure 7.17 Variation of coefficients c_1 and c_2 as a function of applied voltage

for example, a metal to form a small laminated beam or plate. When the piezoceramic layer is actuated, usually by an electric field in the thickness direction of the layer, the dilatational strains in the piezoceramic layer force the beam or plate to bend, much like a bi-metallic beam or plate would respond when the temperature is changed. A subset of bender construction

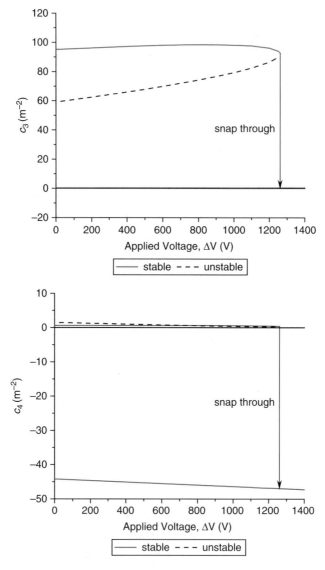

Figure 7.18   Variation of coefficients c_3 and c_4 as a function of applied voltage

uses an elevated-temperature fabrication process either to cure the adhesive bonding the layers together, or to cure the composite layers which are used to replace the metal layers. In either case, when the device is cooled it develops curvatures much the same way as unsymmetric fiber-reinforced composite laminate does. The curvature characteristics for the case of all layers being isotropic are different than when composite layers are used. These small

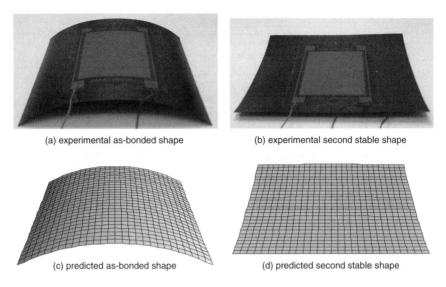

(a) experimental as-bonded shape   (b) experimental second stable shape

(c) predicted as-bonded shape   (d) predicted second stable shape

Figure 7.19 Experimental and predicted stable shapes of actuator–laminate combination with no applied voltage

devices (on the order of 50–200 mm on a side) are used as positioning devices, switches, small pumps, small speakers for noise control, and other applications.

The cross-sections of three such small actuators, referred to as THUNDER™ (Mossi *et al.*, 1998), LIPCA-C1, and LIPCA-C2 (Yoon *et al.*, 2004), are illustrated in Figure 7.20. THUNDER is an acronym for THin layer UNimorph ferroelectric DrivER and sensor, while LIPCA stands for LIghtweight Piezo-composite Curved Actuator. The extensions -C1 and -C2 simply identify different cross-sectional constructions. The thickness of these devices, H in Figure 7.20, is on the order of 0.5–1.0 mm. The LIPCA actuator was introduced to potentially reduce the weight of the THUNDER actuator. As an example of the characteristics of these small unsymmetrically laminated piezoceramic devices, the discussion will focus on THUNDER actuators.

For operation, it is desirable to have the actuators cool from the flat condition at the fabrication temperature to a cylindrical shape, as illustrated in Figure 7.21. With actuation, the curvature is changed to be greater or less than the cooled shape, depending on the sign of the voltage applied. The displacements associated with the changes in curvature are then used for the various applications mentioned above.

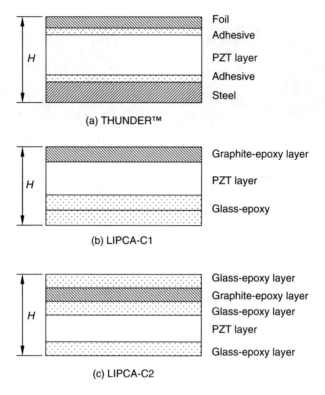

Figure 7.20 Cross-section of several small actuators which are manufactured at an elevated temperature

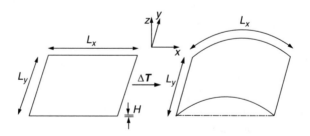

Figure 7.21 Initial and desired cooled shape of actuators and analysis coordinate system

To model the behavior of THUNDER, and LIPCA, actuators, the energy approach and the Rayleigh–Ritz technique were again used (Aimmanee and Hyer, 2004, 2006). The total potential energy included dilatational strains due to thermally induced deformations and actuation of the piezoceramic material. It was similar in form to Equation (7.4) and was given by

$$\Pi = \frac{1}{2} \int_{-\frac{L_x}{2}}^{\frac{L_x}{2}} \int_{-\frac{L_y}{2}}^{\frac{L_y}{2}} \int_{-\frac{H}{2}}^{\frac{H}{2}} \left\{ (\sigma_x - \sigma_x^T - \sigma_x^E)\varepsilon_x + (\sigma_y - \sigma_y^T - \sigma_y^E)\varepsilon_y \right.$$
$$\left. + \left(\sigma_{xy} - \sigma_{xy}^T \right) \gamma_{xy} \right\} dx\,dy\,dz \qquad (7.25)$$

In Equation (7.25)

$$\sigma_x = \overline{Q}_{11}\varepsilon_x + \overline{Q}_{12}\varepsilon_y + \overline{Q}_{16}\gamma_{xy} - \sigma_x^T - \sigma_x^E$$
$$\sigma_y = \overline{Q}_{12}\varepsilon_x + \overline{Q}_{22}\varepsilon_y + \overline{Q}_{26}\gamma_{xy} - \sigma_y^T - \sigma_y^E \qquad (7.26)$$
$$\sigma_{xy} = \overline{Q}_{16}\varepsilon_x + \overline{Q}_{26}\varepsilon_y + \overline{Q}_{66}\gamma_{xy} - \sigma_{xy}^T$$

and Equations (7.6) and (7.7) were used in addition to the terms representing electric field effects in the piezoceramic material, which was assumed to be isotropic, namely,

$$\sigma_x^E = \overline{Q}_{11}\varepsilon_x^E + \overline{Q}_{12}\varepsilon_y^E$$
$$\sigma_y^E = \overline{Q}_{12}\varepsilon_x^E + \overline{Q}_{22}\varepsilon_y^E \qquad (7.27)$$

and

$$\varepsilon_x^E = d_{31}E_3$$
$$\varepsilon_y^E = d_{31}E_3 \qquad (7.28)$$

where d_{31} is the piezoelectric coefficient of the material and the electric field strength through the thickness of piezoceramic material is E_3. The strains and curvatures were assumed to be of the forms given by Equations (7.1), (7.2), and (7.3).

For application of the Rayleigh–Ritz technique, the assumed displacement fields were taken to be

$$u^0(x, y) = c_{10}x + c_{12}x^3 + c_{14}xy^2 + c_{16}x^5 + c_{18}x^3y^2 + c_{20}xy^4 + c_{22}x^7$$
$$v^0(x, y) = c_{11}y + c_{13}y^3 + c_{15}x^2y + c_{17}y^5 + c_{19}x^2y^3 + c_{21}x^4y + c_{23}y^7$$
$$w^0(x, y) = c_1x^2 + c_2y^2 + c_3x^4 + c_4y^4 + c_5x^2y^2 + c_6x^4y^2 + c_7x^2y^4 + c_8x^6 + c_9y^6$$
$$(7.29)$$

As seen, there were 23 coefficients to be solved for, and stability was checked by considering the second variation involving these 23 coefficients.

With the displacement field in Equation (7.29), the curvature was not assumed to be constant. To illustrate the predicted curvatures of the cooled shapes and the effect of actuation on the change in curvatures, the average

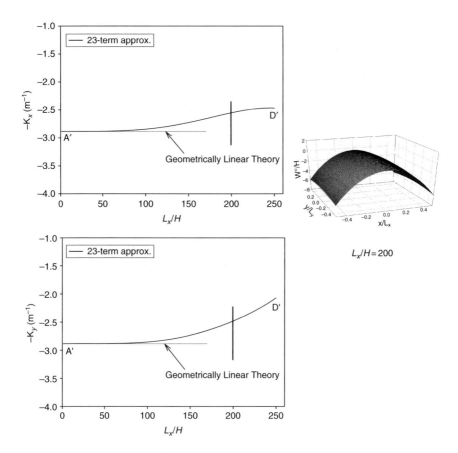

Figure 7.22   Cooled curvatures of beam-like THUNDER™ actuators as a function of L_x/H for $L_y/L_x = 0.33$, and shape for $L_x/H = 200$

curvature was computed as a function of actuator geometry and applied voltage level and will be shown in figures to follow. The average curvatures in the x- and y-directions were defined as

$$K_x = \frac{1}{L_x} \int_{-\frac{L_x}{2}}^{+\frac{L_x}{2}} \kappa_x^0(x, 0)dx$$

$$K_y = \frac{1}{L_y} \int_{-\frac{L_y}{2}}^{+\frac{L_y}{2}} \kappa_y^0(0, y)dy$$

(7.30)

The significant influences of the geometry on the cooled shapes of THUNDER actuators are illustrated in Figures 7.22 and 7.23. The influences on the average curvatures, hereafter referred to as curvatures, of the length-to-thickness ratio, defined as L_x/H and hereafter referred to as length ratio (see Figure 7.21), and whether the actuator is beam-like with an aspect ratio

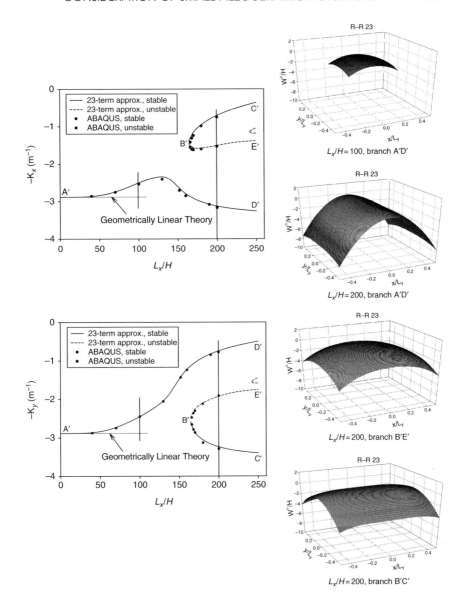

Figure 7.23 Multiple cooled curvatures of plate-like THUNDER™ actuators as a function of L_x/H for $L_y/L_x = 0.70$, and shapes for $L_x/H = 100$ and 200

$L_x/L_y = 0.33$, or plate-like with an aspect ratio $L_x/L_y = 0.70$, are shown in these figures. For an actuator that is beam-like, the influence of the length ratio on the curvatures is illustrated in Figure 7.22. For comparison, the predicted curvatures when geometric nonlinearities, the underlined terms in

Equation (7.2), are ignored are also shown. As can be seen, unlike the results in Figure 7.23 for the plate-like actuator, the predicted curvature vs. length ratio consists of a single branch, identified as branch A′D′. Furthermore, the predictions of the geometrically linear theory are independent of the length ratio and appear to be valid for length ratios up to a value of about 100. The x- and y-direction curvatures are equal, the result being a dome-like shape, which is obscured by the somewhat narrow geometry of the actuator. As the length ratio increases, the curvatures are somewhat suppressed, the y-direction curvature more than the x-direction curvature. The cooled shape for the case of $L_x/H = 200$ is illustrated in the figure and it is clear that the desired cylindrical shape is achieved, though there is noticeable curvature in the y-direction. To be noted is the fact that the curvature in the x-direction appears to be uniform, so the use of the average curvature here to quantify the shape is not that limiting.

When the actuator is plate-like, completely different behavior is predicted, as observed in Figure 7.23. The most noticeable geometric effect is the existence of multiple branches for the curvature vs. length ratio relation. The relation exhibits limit point behavior, at point B′ at a value of $L_x/H = 166$, as opposed to the bifurcation behavior of the antisymmetric cross-ply laminate at point B in Figure 7.2. The primary branch is labeled A′D′, the secondary branch B′C′, and the third branch B′E′. The third branch represents unstable shapes of the actuator. Considering branch A′D′, at values of L_x/H less than 50, geometric nonlinearities are not important. As L_x/H increases beyond 50, both curvatures decrease. At a value of $L_x/H = 100$, the cooled actuator is dome-like with curvatures about 20 % less than predicted by the linear theory. The cooled shape is illustrated in Figure 7.23. A finite-element analysis of the plate-like actuators using the commercial code ABAQUS predicted the average curvatures indicated by the round symbols for the stable shapes and square solutions for the unstable shapes. Needless to say, the agreement is good, particularly the fact that ABAQUS could be made to calculate multiple solutions. Note that the Rayleigh–Ritz approach predicted another branch near point E′. This branch was not pursued with ABAQUS. This branch could be a factor for values of L_x/H near 250.

Continuing on branch A′D′, as L_x/H increases beyond 100, the y-direction curvature continues to decrease, but the x-direction curvature begins to increase. At a value of $L_x/H = 200$ on branch A′D′, the actuator has the desired cylindrical shape, as shown in the figure. The curvature in the x-direction is slightly greater than predicted by the linear theory. Again, there is noticeable curvature in the y-direction.

Considering branch B′C′, the actuator has a cooled shape with considerable curvature in the y-direction and very little in the x-direction, as illustrated in the figure. This shape is not an intended operating shape, but it is predicted to exist. The shape of branch A′D′ can be transformed to the shape of branch B′C′ by a snap-through action. Branch B′E′ represents an actuator with approximately equal curvatures in the x- and y-directions, a dome-like configuration, but such a shape will never be observed due to the lack of stability.

Comparing the results of Figure 7.23 with those of Figure 7.22, it can be concluded that the aspect ratio of the actuator has a significant effect on whether or not multiple cooled shapes exist. From Figure 7.23 it can be concluded that if multiple shapes can exist, the length ratio controls whether the actuator will have multiple shapes or not.

When a voltage is applied to the piezoceramic material, the actuator curvatures change. In Figure 7.24 the curvature vs. length ratio relations for the beam-like actuator with electric field strengths of $\pm 2\,\mathrm{MV}/\mathrm{m}$ applied through the thickness of the piezoceramic material are illustrated. The case of no electric field from Figure 7.22 is included. A positive field causes the curvatures to decrease, while a negative field causes the curvatures to increase. In the geometrically linear range, the magnitudes of the changes in curvature due a positive field are the same as the magnitudes of the changes in curvature due to a negative field. However, as L_x/H increases beyond 100, this is not the case. At $L_x/H = 200$, for example, the magnitude of the positive curvature change caused by $+2\,\mathrm{MV}/\mathrm{m}$ is slightly larger than the magnitude of the negative curvature change caused by $-2\,\mathrm{MV}/\mathrm{m}$. Thus an electric field that is harmonic in time would not result in displacements that are harmonic in time.

The electric-field-induced curvature vs. length ratio relations for the plate-like actuator are shown in Figure 7.25, along with the zero voltage case. For the primary branch, A′D′, a positive electric field causes curvatures to decrease in both the x- and y-directions, as occurred with the beam-like actuators. However, for $L_x/H = 200$ the change in y-direction curvature is very small, so the actuator would be observed to change curvature only in the x-direction, the intended mode of operation. It is important to observe that the changes in the x-direction curvature for the $\pm 2\,\mathrm{MV}/\mathrm{m}$ electric field strengths are not uniform with L_x/H. For L_x/H approximately 115 to 120 the changes in x-direction curvature due to the applied electric field are minimal compared with other values of L_x/H, as determined by the vertical distances between the curvatures of the cooled actuator and the curvatures for $\pm 2\,\mathrm{MV}/\mathrm{m}$. For $L_x/H = 200$ the changes in curvature are among the largest of those computed. The dependence of the change in curvature on L_x/H is due to geometric nonlinearities and illustrates that indeed geometrically nonlinear effects can enhance the curvature changes. On the other hand, they can suppress curvature changes. It should be noted that the limit point shifts to the left or right due to when voltage is applied, moving to the left

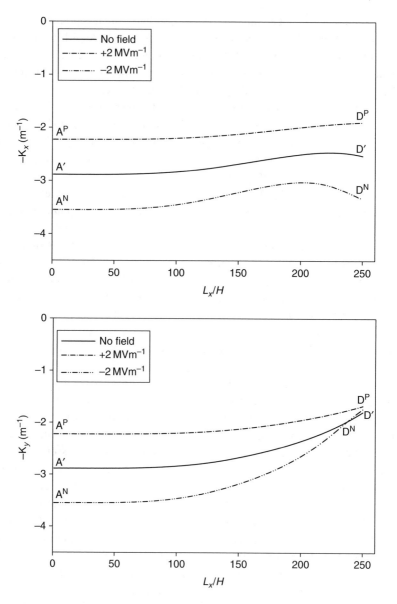

Figure 7.24 Effect of actuation on curvatures of beam-like THUNDER™ actuators, $L_y/L_x = 0.33$

with a positive voltage and to the right with a negative voltage. An actuator constructed to have a value of L_x/H near the limit point value of 166 could exhibit unusual behavior when energized with a voltage that continually changes sign, such as a harmonic voltage time history.

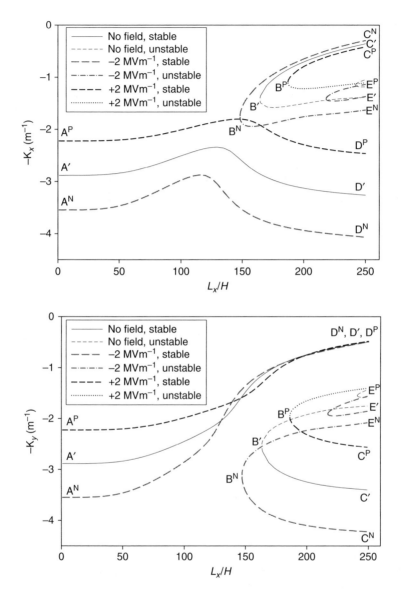

Figure 7.25 Effect of actuation on curvatures of beam-like THUNDER™ actuators, $L_y/L_x = 0.70$

Before closing, it should be mentioned that geometric nonlinearities can have a significant influence on the behavior of circular actuators (Hyer and Jilani, 2000, 2002, 2003).

7.8   CONCLUSIONS

It can be concluded that geometric nonlinearities have considerable impact on the predicted deformation behavior of the various laminated plates and small devices discussed. Geometric nonlinearities lead to a sensitivity of the deformation characteristics to geometric parameters that is not present when geometric nonlinearities are ignored. The existence of multiple shapes is also an important effect of geometric nonlinearities. This characteristic can be used to advantage, or it can be a disadvantage that should be avoided, depending on the application. An interesting observation is that while the layered composite plates studied resulted in multiple shapes with curvatures of opposite sign, the layered isotropic plates resulted in multiple shapes with curvatures of the same sign.

REFERENCES

Aimmanee, S. and M.W. Hyer (2004), 'Analysis of the Manufactured Shape of Rectangular THUNDER-Type Actuators,' *Smart Materials and Structures*, 13, 1389–1406

Aimmanee, S. and M.W. Hyer (2006), 'A Comparison of Various Piezoceramic Actuators,' *Journal of Intelligent Material Systems and Structures*, 17(2), 167–86

Dano, M.-L. and M.W. Hyer (1996), 'The Response of Unsymmetric Laminates to Simple Applied Forces,' *Mechanics of Composite Materials and Structures*, 3, 65–80

Dano, M.-L. and M.W. Hyer (1998), ' Thermally-Induced Deformation Behavior of Unsymmetric Laminates,' *International Journal of Solids and Structures*, 35(17), 2101–20

Dano, M.-L. and M.W. Hyer (2002), 'Snap Through Behavior of Unsymmetric Composite Laminates,' *International Journal of Solids and Structures*, 39(1), 175–98

Dano, M.-L. and M.W. Hyer (2003), 'SMA-Induced Snap-Through of Unsymmetric Fiber-Reinforced Composite Plates,' *International Journal of Solids and Structures*, 40, 5949–72

Hamamoto, A. and M.W. Hyer (1987), 'Nonlinear Temperature-Curvature Relationships for Unsymmetric Graphite-Epoxy Laminates,' *International Journal of Solids and Structures*, 23(7), 919–35

Hyer, M.W. (1981a), 'Some Observations on the Cured Shapes of Thin Unsymmetric Laminates,' *Journal of Composite Materials*, 15, 175–94

Hyer, M.W. (1981b), 'Calculations of the Room-Temperature Shapes of Unsymmetric Laminates,' *Journal of Composite Materials*, 15, 296–310

Hyer, M.W. (1982), 'The Room-Temperature Shapes of Four-Layer Unsymmetric Cross-Ply Laminates,' *Journal of Composite Materials*, 16, 318–40

Hyer, M.W. (1998), *Stress Analysis of Fiber-Reinforced Composite Materials*, WCB/McGraw-Hill, New York

Hyer, M.W. and A.B. Jilani (2000), 'Predicting the Axisymmetric Manufacturing Deformations of Disk- Style Benders,' *Journal of Intelligent Material Systems and Structures* (Special edition from 4th ARO Workshop on Smart Structures), 11(5), pp. 370–81

Hyer, M.W. and A.B. Jilani (2002), 'Deformation Characteristics of Circular RAINBOW Actuators,' *Smart Materials and Structures,* 11, 175–95, 2002

Hyer, M.W. and A.B. Jilani (2003), 'Manufactured Configurations of Piezoceramic Disk-Style Actuators,' *Journal of Intelligent Material Systems and Structures,* 14(6), 359–70

Mossi, K.M., G.V. Shelby, and R.G. Bryant (1998), 'Thin-layer Composite Unimorph Ferroelectric Driver and Sensor Properties,' *Materials Letters,* 35(1/2), 39–49

Schlecht, M., K. Schulte, and M.W. Hyer (1995), 'Advanced Calculations of the Room-Temperature Shapes of Thin Unsymmetric Composite Laminates,' *Composite Structures,* 32, 627–33

Schultz, M.R. and M.W. Hyer (2003), 'Snap-Through of Unsymmetric Cross-Ply Laminates using Piezoceramic Actuators,' *Journal of Intelligent Material Systems and Structures,* 14, 795–814

Wilkie, W.K., R.G. Bryant, J.W. High, R.L. Fox, R.F. Hellbaum, A. Jalink, Jr., B.D. Little, and P.H. Mirick (2000), 'Low-Cost Piezocomposite Actuator for Structural Control Applications,' 7th Annual International Symposium on Smart Structures and Materials, Newport Beach, CA (available International Society for Optical Engineering (SPIE), PO Box 10, Bellingham, WA 98227, USA)

Williams, R.B., M.R. Schultz, M.W. Hyer, D.J. Inman, and W. Keats Wilkie (2004), 'Nonlinear Tensile and Shear Behavior of Macro Fiber Composite Actuators,' *Journal of Composite Materials,* 38(10), 855–69

Yoon, K.J., K.H. Park, S.K. Lee, N.S. Goo, and H.C. Park (2004), 'Analytical Design Model for a Piezo-Composite Unimorph Actuator and its Verification using Lightweight Piezo-Composite Curved Actuators,' *Smart Materials and Structures,* 13, 459–67

8

Negative Stiffness and Negative Poisson's Ratio in Materials which Undergo a Phase Transformation

T.M. Jaglinski[1] and R.S. Lakes[2]

[1] *Institute for Shock Physics, Washington State University, PO Box 642816, Pullman, WA 99164–2816, USA*
[2] *University of Wisconsin, 541 Engineering Research Building, 1500 Engineering Drive, Madison, WI 53706–1687, USA*

8.1 INTRODUCTION

For most elastic systems, stiffness is positive. This means that a deformed object experiences a force in the same direction as the deformation. Negative stiffness can occur in systems such as pre-strained objects including post-buckled elements, which contain stored energy[1]. Heterogeneous systems with one constituent of negative stiffness are predicted to give rise to high damping and stiffness[2], higher than that of either constituent. Experimentally, high viscoelastic damping and negative axial stiffness were observed[3] in lumped systems containing post-buckled viscoelastic rubber tubes. High viscoelastic damping has also been observed in metal matrix (Sn) composites containing

Adaptive Structures: Engineering Applications   Edited by D. Wagg, I. Bond, P. Weaver and M. Friswell
© 2007 John Wiley & Sons, Ltd

particulate ferroelastic VO_2 inclusions[4]. Ferroelastics are of interest in this context since they are predicted if constrained, to exhibit negative stiffness below the transformation temperature. Ferroelastics, if not constrained, undergo a shear instability in which the crystal structure changes form as temperature is lowered. In the Landau theory of phase transformations, the free energy has a relative maximum, corresponding to unstable equilibrium, below the material's transformation temperature T_c[5]. Since the curvature of this energy function corresponds to a modulus, negative moduli are predicted. These negative moduli are not observed in blocks free of constraint since the material forms bands or domains due to the instability. Multi-domain blocks of material have positive stiffness as is observed experimentally.

The continuum theory of elasticity provides guidance as to the kinds of instability to be expected. In isotropic elastic solids, the 'allowable' range of Poisson's ratio v is

$$-1 < v < 0.5. \tag{8.1}$$

This corresponds to the requirement that the shear (G) and bulk (K) moduli[6] be positive for *stability* of an unconstrained block of material. This condition corresponds to a surface traction boundary condition in elasticity theory; the condition means that the surface forces are specified. Most isotropic materials have Poisson's ratios close to 0.3; rubbery materials have Poisson's ratios close to 0.5. Negative Poisson's ratio, though counter-intuitive, is possible. Foams[7][8] with v as small as -0.8 have been conceptualized, fabricated, and studied. The composites of Milton[9] offer the intriguing possibility of stiff materials with a negative Poisson's ratio. The stiffness of one of the two constituents must be at least 25 times greater than that of the other constituent to obtain a Poisson's ratio less than zero. A Poisson's ratio approaching the stability limit of -1 requires constituents which differ even more in stiffnesses: one phase becomes very soft, tending to 'empty space' in its properties. Microporous polymers with negative Poisson's ratio have been developed [10]. Another intriguing possibility is that of creating solid polymer materials with negative Poisson's ratio by design on the molecular scale[11].

The range of properties associated with stability of a *constrained* object is that of strong ellipticity. The range of elastic constants is[12], for isotropic solids,

$$G > 0 \tag{8.2a}$$

and

$$v < 0.5 \text{ or } v > 1. \tag{8.2b}$$

Since $E = 2G(1 + v)$, and $K = 2G(1 + v)/3(1 - 2v)$, negative Young's modulus E and bulk modulus K are allowed. Specifically

$$-\infty < E < \infty \text{ or } -4G/3 < K < \infty. \tag{8.3}$$

Negative Poisson's ratios below the usual stability limit of -1 are possible. Strongly elliptic materials are stable with respect to the formation of domains or bands. An inclusion in a composite is partially constrained. Recently it has been shown that the range of material properties associated with such a situation is intermediate between positive definiteness and strong ellipticity [13].

In viscoelastic materials, one may express the behavior in the context of the dynamic viscoelastic moduli. The complex dynamic Young's modulus is $E^* = E' + iE'' = E'(1 + i \tan \delta)$, with $E' = \text{Re}\{E^*\}$ and $\tan \delta \equiv \text{Im}\{E^*\}/\text{Re}\{E^*\}$; δ is the phase angle between the stress and strain sinusoids. The primes are standard notation for the real and imaginary part respectively; they do not represent derivatives. The dynamic modulus is a function of frequency. In composite materials it depends on constituent properties and morphology. A representative stiffness-loss map for well-known materials is shown in Figure 8.1. The product $E' \tan \delta$ is a figure of merit [14] for the damping of structural vibration. Most known materials [15], including

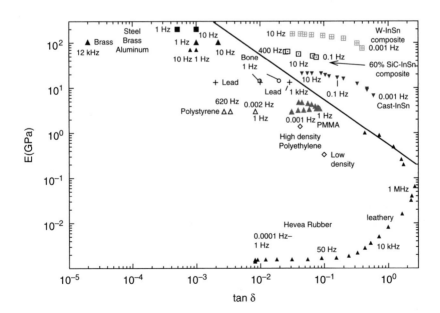

Figure 8.1 Stiffness-loss map: properties of selected materials including designed particulate composites

commercial damping layers, exhibit maximum E' tan $\delta < 0.6\,$GPa. Some composites with values as much as an order of magnitude higher have been developed[16][17]; selected results are also shown in Figure 8.1.

In the present work we explore the role of negative stiffness constituents in composites and constrained negative stiffness in the grains of polycrystals in achieving high damping in materials.

8.2   EXPERIMENTAL METHODS

8.2.1   Material Preparation

For InTl alloy, indium wire (Alfa Aesar, 99.9985 % metals basis) and thallium granules (Alfa Aesar, 99.999 % metals basis) were weighed, mixed, and cast in a steel mould to form a In–21at%–Tl alloy. The mould cavity was rectangular (8.0 cm long, 1.9 cm wide, and 1.5 cm deep) and was provided with tapered surfaces to allow easy removal of the ingot. Specimens were cut using a low-speed abrasive saw from the cast ingot into rectangular cross-sections, typically on the order of $3 \times 4 \times 30\,$mm. Specimens were cut and polished and observed using light microscopy.

Solid rods of barium titanate ceramic were obtained from two sources. Prof. Smay, of Oklahoma State University, synthesized a cylindrical rod 73 cm long and 21 cm in diameter by sintering from Ticon-HPB powder (Ferro Electronic Materials, Niagara Falls, NY) at 1350 °C for 1 hour. Test-bourne Ltd. (Hampshire, UK) provided five rods about 6 cm long and 12 cm in diameter. These as-received rods had tapered ends, which were sectioned off using a diamond saw. The ends were then polished to achieve approximate parallelism.

Viscoelastic properties of specimens of InTl and of composites were determined in torsion or bending using broadband viscoelastic spectroscopy (BVS)[18]. The instrument was provided with temperature control and the ability to measure slow, large-amplitude, free-end deformation associated with spontaneous strain. The method is capable of frequency-dependent measurements over 11 orders of magnitude of frequency to about 100 kHz. In the present studies, temperature dependence is emphasized since temperature allows one to tune the behavior of ferroelastic solids. Torque on the end of the specimen was generated via the action of a Helmholtz coil on a high-intensity magnet cemented to the end. The fixed end of the specimen was clamped to a 25 mm diameter steel support rod by tungsten adapters and set screws. Angular displacement of the free end was determined by reflecting a laser beam from a mirror cemented to a drive magnet, upon a split diode light sensor. The sensor electrical signal was input to a lock-in amplifier: viscoelastic damping and modulus measurements were taken using a Stanford Research Systems lock-in amplifier (SR850 DSP) by measuring the phase angle between the sinusoidal torque and displacement signals. Constant frequency tests with variable temperature were conducted at various frequencies well below the specimen resonance (typically about 2 kHz in torsion). Spontaneous strain and other manifestations of instability were studied by splitting a portion of the laser beam and directing it to a wide-angle, two-axis photodiode position sensor (Pacific Silicon Sensor Inc. DL100-7PCBA, Westlake, CA) with a detector area of $1 \, cm^2$. Specimen temperature was measured using a type-K (Omega) thermocouple about 1 mm from the base of the specimen in the stream of heated air. Output voltage was recorded by a digital oscilloscope.

Polycrystalline ceramic $BaTiO_3$ was studied using pulsed wave ultrasound, and by mechanical testing. Both longitudinal and shear wave transit times were measured ultrasonically at 1 MHz, and engineering moduli were calculated using the theory of isotropic elasticity. After ultrasonic testing, the rods were provided with SK-09-125TM-350 strain gauge T-rosettes using standard M-bond-610 strain gauge cement. Thermal expansion tests were done by heating and cooling the rods in a furnace, and monitoring the strain. Specimen temperature was measured using a type-K (Omega) thermocouple in contact with the specimen. Strain results were corrected for expansion mismatch between gauge and ceramic. A differential scanning calorimetry (DSC) test was done on granules of barium titanate obtained from Alfa Aesar. Some granules were plated with nickel for later use as inclusions in composites; these were also examined by DSC. Compression tests upon solid rods of barium titanate were conducted on a MTS (Minneapolis, MN) 20 000 lb (9050 kg) capacity servo-hydraulic frame. To achieve thermal isolation, aluminum oxide rods 2.54 cm in diameter and 10 cm long were placed between the specimen and the steel test frame platens; also, a swivel joint was added between the actuator and the lower alumina isolation rod. As above, phase and magnitude measurements were conducted using the Stanford Research Systems lock-in amplifier (SR850 DSP). Mechanical testing using this frame was done at 1 Hz. Phase calibration was done with 6061 aluminum alloy which has a tan δ on the order of 10^{-5}. Based on measured phase for this alloy, a phase shift correction of about 0.01 rad at 1 Hz was applied to compensate for overall shifts in the servo-hydraulic system electronics.

8.3 COMPOSITES

8.3.1 Theory

In composite materials there are several exact analytical solutions for composite properties in terms of inclusion properties and geometry. Within these solutions one can allow the modulus of an inclusion to become negative and calculate the composite properties. If the matrix is much stiffer than the inclusion, the situation approximates that of constraint of surface displacement in which strong ellipticity suffices for stability. This allows negative bulk modulus. Negative shear moduli are excluded. In real materials, domains form at temperatures below the point at which a shear modulus softens to zero. In crystals which are sufficiently small, surface energy, not incorporated in the continuum theory, results in single domain single crystals.

The Voigt formula, corresponding to fibers or laminae aligned in the direction of a uniaxial stress, is

$$E_c = E_1 V_1 + E_2 V_2, \tag{8.4}$$

in which E_c, E_1, and E_2 refer to the Young's modulus (stiffness) of the composite, phase 1 and phase 2, and V_1 and V_2 refer to the volume fraction of phase 1 and phase 2 with $V_1 + V_2 = 1$. Negative stiffness of one phase can give rise to a small or zero composite stiffness.

The Reuss formula is, in terms of the compliances, e.g., $J_1 = 1/E_1$,

$$J_c = J_1 V_1 + J_2 V_2. \tag{8.5}$$

Negative stiffness of one phase can give rise to a large composite stiffness, even a stiffness which tends to infinity. The corresponding composite geometry consists of laminae orthogonal to the direction of a uniaxial stress. This configuration is unstable under load control but it can be stabilized in displacement control for a range of constituent properties. The Voigt and Reuss formulae represent bounds provided that both phases are described by a positive definite strain energy. Although this excludes negative stiffness for interpretation of formulae as bounds, they remain exact analytical solutions for the corresponding composite morphologies when constituents assume negative moduli.

The Hashin–Shtrikman formulae[19] apply for isotropic composites. They are bounds provided that both phases are described by a positive definite strain energy, which excludes negative stiffness. The lower bound for the shear modulus G_L of an *elastic* composite is

$$G_L = G_2 + \cfrac{V_1}{\cfrac{1}{G_1 - G_2} + \cfrac{6(K_2 + 2G_2)V_2}{5(3K_2 + 4G_2)G_2}}, \qquad (8.6)$$

in which K_1, K_2, G_1 and V_1, and G_2 and V_2, are the bulk modulus, shear modulus, and volume fraction of phases 1, and 2, respectively. Both shear and bulk moduli must be positive for this and related formulae to be bounds[20]. Upper and lower Hashin–Shtrikman formulae for bulk modulus are attained via a hierarchical morphology in which the composite is filled with coated spheres of different size but identical ratio of sphere size to coating thickness. The attainment is *exact* for the bulk modulus[21] and approximate for the shear modulus. The shear modulus formula is attained *exactly* by hierarchical laminates[22]. Since they are exact solutions, they may be used in predictions of behavior of composites with negative stiffness constituents. For viscoelastic solids, the moduli or compliances in the above equations become complex following the elastic–viscoelastic correspondence principle. If negative moduli are allowed in these formulae, mechanical damping of the composite can approach a singularity; sigmoid-shaped anomalies are predicted in the composite moduli as a function of inclusion modulus.

8.3.2 Experiment

High-damping composite materials can be achieved in several ways. The simplest approach is to embed inclusions of a stiff, low-damping phase in a matrix of more compliant high-damping material. The optimal inclusion shape, considering various inclusion morphologies, is a sphere of stiff material coated with a layer of high-damping material forming the hierarchical morphology referred to above[15]. Spherical or near spherical inclusions of similar size will suffice for most purposes. If anisotropy is tolerable, a Reuss laminate provides better results in this context than a Voigt laminate. Representative results of this approach are shown in the map in Figure 7.1. Negative stiffness or modulus allows larger damping, even damping tending to infinity, to be achieved, as predicted theoretically[2] via the above equations. Experimental realization was achieved in a lumped system involving buckled tubes[3]; damping increases of orders of magnitude were observed. As for distributed systems, several experimental studies have been conducted in the Sn–VO$_2$ system[4] [23] [24]. Vanadium dioxide is a ferroelastic material; it undergoes a phase transformation from monoclinic to tetragonal at $T_c = 67\,°\text{C}$. The domain size is typically about $10\,\mu\text{m}$. The rationale is that in single domains of ferroelastic material, domain formation is suppressed by surface energy considerations. Therefore single domain particles, when constrained by a sufficiently stiff composite matrix, can be stable. Inclusions were prepared with a particle size distribution including granules of size smaller than $10\,\mu\text{m}$. Composites with $1\,\%$ volume fraction of inclusions, a value close to the calculated stability limit, exhibited large peaks in mechanical damping (tan δ), and large anomalies in stiffness. Anomalies in stiffness and damping in the composites are much larger than they could be for inclusions of any positive stiffness or damping. Composites with higher concentrations of inclusions exhibited multiple peaks in damping, and manifestations of instability including episodes of negative damping and a thrashing instability. Current experiments as of this writing deal with barium titanate as an inclusion. The experiments upon bulk barium titanate described below are motivated in part by the desire to characterize the material for inclusions. The role of negative stiffness of crystallite grains in polycrystalline material is also explored.

8.4 POLYCRYSTALS

8.4.1 Theory

The rationale for existence of negative stiffness in polycrystals is as follows. Many materials in the vicinity of a solid to solid phase transformation display a decrease (softening) to zero of one or more of the elastic constants. For example, in single crystal InTl, the $(1/2)(C_{11} - C_{12})$ shear modulus approaches zero[25] [26] as the martensitic transformation temperature is approached during heating or cooling. In polycrystals, a grain or crystal which has a negative modulus can be constrained, hence stabilized, by the surrounding grains. This situation is similar to a negative modulus inclusion in a composite; the difference is that composite inclusion concentration can be made dilute to achieve stability, but in a polycrystal the concentration of grains is $100\,\%$.

Polycrystal properties may be predicted using bounding formulae. Upper, G_U, and lower, G_L, bounds[27] for cubic materials are given by

$$G_L = G_1 + 3 \left(\frac{5}{G_2 - G_1} + 4 \frac{3}{5G_1} \frac{K + 2G_1}{3K + 4G_1} \right)^{-1}, \tag{8.7}$$

$$G_U = G_2 + 2 \left(\frac{5}{G_2 - G_1} + 6 \frac{3}{5G_2} \frac{K + 2G_2}{3K + 4G_2} \right)^{-1}. \qquad (8.8)$$

Similar analyses have also been conducted for materials of lower symmetry[28].

Assuming the shear modulus G_1 to become negative provides several predictions for the polycrystalline behavior. The upper Hashin–Shtrikman formula, Equation 8.8, predicts a singularity in internal friction and a sigmoid-shaped curve for the polycrystalline shear modulus.

As for the bulk modulus of the grains, for cubic symmetry, the bulk modulus of the polycrystalline aggregate is the same as that of the crystals.

Consider modulus elements $C_{11} = 1$, $C_{12} = 1$, $C_{44} = 1(1 + 0.05i)$. Then, in the cubic system, the bulk modulus is $K = (C_{11} - 2C_{12})/3$. Suppose the temperature scale is normalized so that the transformation temperature T is one in normalized units. Then let the bulk modulus soften according to

$$K = [(C_{11} - 2C_{12})/3][1000(T - 1)]$$

and can become negative below T. Suppose further that the shear modulus G_1 above softens according to

$$G_1 = 0.2 \, [460|T - 1|][1 + 10^{-5}|T - 1|^{-2}] + 1.$$

Here the shear modulus softens, but not to zero, and is not allowed to go negative, since the crystal grains are assumed to be sufficiently large that many domains form within them. An imaginary, temperature-dependent part is allowed for the modulus to account for viscoelastic dissipation. Results of this analysis are shown in Figure 8.2. There is a peak in mechanical damping as the material is cooled, then damping becomes negative, corresponding to instability, hence spontaneous strain. In considering this as a model for barium titanate, a free crystal of material undergoes a phase transformation from cubic to orthorhombic at the Curie point. In a polycrystalline ceramic or in a composite, there is some constraint on each crystal. Therefore the crystal will remain cubic over some range of temperature below the Curie point, so the analysis is likely to be applicable over a range of temperature down to and somewhat below the Curie point.

8.4.2 Experimental Results

Results are presented for polycrystalline ceramic $BaTiO_3$; for comparison, selected results for polycrystalline InTl are presented as well. Further details

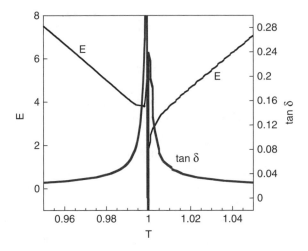

Figure 8.2 Modulus (arbitrary units) and damping predicted in a polycrystalline solid in which the bulk modulus softens to zero at normalized temperature $T = 1$, then to negative values, and in which the shear modulus softens to a positive value

regarding the InTl results and their interpretation are provided elsewhere [23]. Viscoelastic damping tan δ at 100 Hz of In–21 at%–Tl alloy during cooling is shown in Figure 8.3. Shown for comparison are single crystal results from Li *et al.* [29]. Figure 8.4 shows the shear modulus and damping for polycrystalline InTl at two different cooling rates.

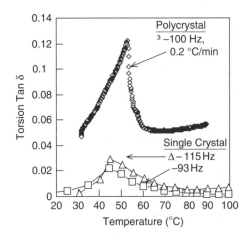

Figure 8.3 Damping peak during cooling at 100 Hz for polycrystalline InTl in comparison to single crystals of similar composition from Li *et al.*, after Ref. [23]

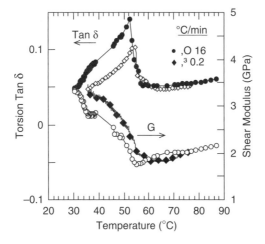

Figure 8.4   Modulus and damping during cooling at 100 Hz for polycrystalline InTl for several cooling rates, after Ref.[23]

The large peak in the tan δ of polycrystalline InTl differs from the usual internal friction peaks in polycrystals, which are broader and weaker than those of single crystals. The usual broadening of damping peaks in polycrystalline solids is attributed to a superposition of contributions from different crystals to the viscoelastic behavior[30]. Martensitic bands occur in both polycrystals and single crystals, so interface theories cannot readily account for the higher peak. Moreover, the high-temperature portion of the damping peak occurs above the temperature at which martensitic bands are observed, therefore this portion cannot be due to the motion of interfaces such as twin boundaries. Spontaneous strain, a manifestation of instability, is shown in Figure 8.5. Constrained negative moduli of the constituent crystallites can account for this damping, as well as for amplification of viscoelastic damping peaks in these polycrystals. Sigmoid-shaped anomalies in the shear modulus vs. temperature at high cooling rates are not predicted by prior analyses of phase transformation but are predicted by analysis of polycrystal properties allowing negative moduli of individual crystallites.

As for barium titanate, room temperature ultrasonic testing at 1 MHz revealed the tensorial C_{11} modulus and shear modulus. From these it was calculated that the thin rods from Testbourne had a Young's modulus of about 100 GPa, a shear modulus of 38 GPa and a Poisson's ratio of 0.34. The 'normal' value of Poisson's ratio supports the notion of isotropy for the polycrystalline specimens. The thicker rod was somewhat less stiff, with E of about 76 GPa, G of 26 GPa, and a Poisson's ratio of about 0.3. The difference

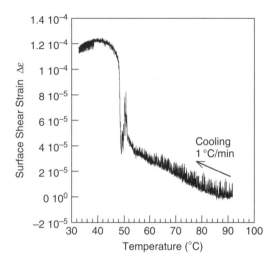

Figure 8.5   Spontaneous strain in polycrystalline InTl, after Ref. [23]

is attributed to imperfect consolidation during preparation of the ceramics. Indeed, the thinner rods had a density of $5.8 \, g/cm^3$, and the thicker rod had a density of $5.54 \, g/cm^3$. By contrast, the accepted density for $BaTiO_3$ in the absence of porosity is $5.85 \, g/cm^3$.

Thermal deformation results for barium titanate are shown in Figure 8.6. A spontaneous strain was observed near the Curie point transformation

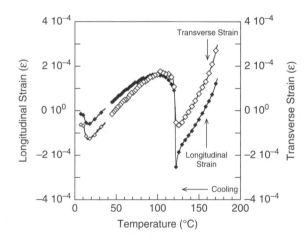

Figure 8.6   Free thermal deformation during cooling for polycrystalline barium titanate showing spontaneous strain for both the cubic to tetragonal transition near 125 °C, and the tetragonal to orthorhombic transition near 15 °C for this material

temperature of 120 °C during cooling. The corresponding jump in strain during heating occurred near 140 °C. Such hysteresis is known in this system. The hysteresis is also manifested in DSC as shown in Figure 8.7. Viscoelastic behavior of barium titanate in load control is shown in Figure 8.8. The damping peak associated with phase transformation is typical. The damping

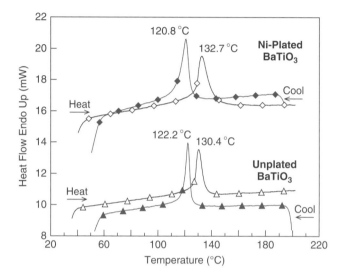

Figure 8.7 Differential scanning calorimetry (DSC) of polycrystalline barium titanate. Ends refer to endothermic

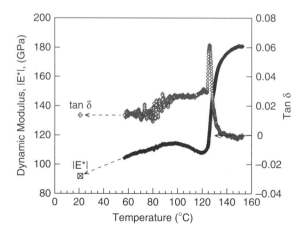

Figure 8.8 Young's modulus and tan δ for polycrystalline barium titanate at 1 Hz as a function of temperature

peak begins and reaches its maximum at a temperature well above the corresponding temperatures associated with the instability of spontaneous strain and with the peak in the DSC curve. As with the InTl, the peak in the barium titanate damping appears in a regime where there are no domain wall interfaces. A role for negative modulus of crystallites is therefore suggested in this case as well. A moderate softening of modulus occurs below the Curie point. The ripples in properties near 60 °C are notable. We hypothesize that they represent an effect of constraint by surrounding grains upon the transformation of crystallite grains.

8.5   DISCUSSION

For adaptive structures, the spontaneous strain associated with phase transformations can be of interest, not only in shape memory materials, but also in actuators and ceramics. Snap-through can be done via macroscopic or composite structure, e.g., tape springs[31]; alternatively it can be done in connection with phase transformations in composites. In the negative stiffness paradigm, interface deformation is amplified via the balance between positive and negative stiffness. This interface may be between inclusion and matrix in a composite or between adjacent crystallites in a polycrystalline metal or ceramic. As for composites, the spontaneous volumetric strain which occurs in the free expansion of the ceramic offers the potential that negative bulk modulus may occur if this transformation is subjected to volumetric constraint. Among the many possible areas of application are high-damping layers for vibration abatement, and materials with enhanced piezoelectric or thermal[32] properties.

REFERENCES

[1] Thompson, J.M.T. (1979). Stability prediction through a succession of folds. *Philos. Trans. R. Soc.*, **292**: 1–23.

[2] Lakes, R.S. (2001). Extreme damping in composite materials with a negative stiffness phase. *Phys. Rev. Lett.*, **86**: 2897–2900.

[3] Lakes, R.S. (2001). Extreme damping in compliant composites with a negative stiffness phase. *Philos. Mag. Lett.*, **81**: 95–100.

[4] Lakes, R.S., Lee, T., Bersie, A., and Wang, Y.C. (2001). Extreme damping in composite materials with negative stiffness inclusions. *Nature*, **410**: 565–567.

[5] Falk, F. (1980). Model free energy, mechanics, and thermodynamics of shape memory alloys. *Acta Metall.*, **28**: 1773–1780.

[6] Timoshenko, S.P. and Goodier, J.N. (1970). *Theory of Elasticity*, McGraw-Hill, New York, 3rd edition.

[7] Lakes, R.S. (1987). Foam structures with a negative Poisson's ratio. *Science*, **235**: 1038–1040.

[8] Lakes, R.S. (1993). Advances in negative Poisson's ratio materials. *Adv. Mater.*, **5**: 293–296.

[9] Milton, G. (1992). Composite materials with Poisson's ratios close to −1. *J. Mech. Phys. Solids*, **40**: 1105–1137.

[10] Caddock, B.D. and Evans, K.E. (1989). Microporous materials with negative Poisson's ratios. I. Microstructure and mechanical properties. *J. Phys. D: Appl. Phys.*, **22**(12): 1877–1882.

[11] Evans, K.E., Nkansah, M.A., Hutchinson, I.J., and Roger, S.C. (1991). Molecular network design. *Nature*, **353**: 124.

[12] Knowles, J.K. and Sternberg, E. (1978). On the failure of ellipticity and the emergence of discontinuous gradients in plane finite elastostatics. *J. Elasticity*, **8**: 329–379.

[13] Drugan, W. (2007). Elastic composite materials having a negative stiffness can be stable. *Phys. Rev. Lett.*, **98**, 055502.

[14] Kerwin, E.M. Jr. and Ungar, E.E. (1990). Requirements imposed on polymeric materials in structural damping applications. In *Sound and Vibration Damping with Polymers*, ed. R.D. Corsaro and L.H. Sperling, American Chemical Society, Washington, DC.

[15] Chen, C.P. and Lakes, R.S. (1993). Analysis of high loss viscoelastic composites. *J. Mater. Sci.*, **28**: 4299–4304.

[16] Ludwigson, M., Swan, C.C. and Lakes, R.S. (2002). Damping and stiffness of particulate SiC-InSn composite. *J. Compos. Mater.*, **36**: 2245–2254.

[17] Brodt, M. and Lakes, R.S. (1995). Composite materials which exhibit high stiffness and high viscoelastic damping. *J. Compos. Mater.*, **29**: 1823–1833.

[18] Lee, T., Lakes, R.S. and Lal, A. (2000). Resonant ultrasound spectroscopy for measurement of mechanical damping: comparison with broadband viscoelastic spectroscopy. *Rev. Sci. Instrum.*, **71**: 2855–2861.

[19] Hashin, Z. and Shtrikman, S. (1963). A variational approach to the theory of the elastic behavior of multiphase materials. *J. Mech. Phys. Solids*, **11**, 127–140.

[20] Gibiansky L.V. and Torquato, S. (1993). Link between the conductivity and elastic moduli of composite materials. *Phys. Rev. Lett.*, **71**: 2927–2930.

[21] Hashin, Z. (1962). The elastic moduli of heterogeneous materials. *J. Appl. Mech.*, **29**: 143–150.

[22] Milton, G.W. (1986). Modelling the properties of composites by laminates. In *Homogenization and effective moduli of materials and media*, ed. J.L. Erickson, D. Kinderlehrer, R. Kohn, J.L. Lions, (springer Verlag, Berlin, pp. 150–175.

[23] Jaglinski, T. Frascone, P., Moore, B., Stone, D., and Lakes, R.S. (2006). Internal friction due to negative stiffness in the indium-thallium martensitic phase transformation. *Philos. Mag.*, **86**: 4285–4303.

[24] Wang, Y.C., Ludwigson, M., and Lakes, R.S. (2004). Deformation of extreme viscoelastic metals and composites. *Mater. Sci. Eng. A*, **370**: 41–49.

[25] Gunton, D.J. and Saunders, G.A. (1974). The elastic behaviour of In-Tl alloys in the vicinity of the martensitic transition. Solid State Commun., **14**: 865–868.

[26] Pace, N.G. and Saunders, G.A. (1972). Ultrasonic study of lattice stability in indium + thallium alloys. *Proc. R. Soc. A*, **326**: 521–533.

[27] Hashin, Z. and Shtrikman, S. (1962). A variational approach to the theory of the elastic behaviour of polycrystals. *J. Mech. Phys. Solids*, **10**: 343–352.

[28] Watt, J.P. and Peselnick, L. (1980). Clarification of the Hashin-Shtrikman bounds on the effective elastic modulii of polycrystals with hexagonal, trigonal, and tetragonal symmetries. *J. Appl. Phys.*, **51**(3): 1525–1531.

[29] Li, J. Zhou, X. and Wuttig, M. (1990). Internal friction in In-Tl alloys. *Scr. Metall. Mater.*, **24**: 901–902.

[30] Nowick, A.S. and Berry, B.S. (1972). *Anelastic Relaxation in Crystalline Solids*, Academic press, New York, 435–462.

[31] Pellegrino, S. and Seffen, K.A. (1999). Deployment dynamics of tape springs. *Proc. R. Soc. A*, **455**: 1003–1048.

[32] Wang, Y.C. and Lakes, R.S. (2001). Extreme thermal expansion, piezoelectricity, and other coupled field properties in composites with a negative stiffness phase. *J. Appl. Phys.*, **90**: 6458–6465.

9
Recent Advances in Self-Healing Materials Systems

M.W. Keller[1,2], B.J. Blaiszik[1,2], S.R. White[2,3] and N.R. Sottos[2,4]

[1]*Theoretical and Applied Mechanics Program, Department of Mechanical Science and Engineering, University of Illinois Urbana–Champaign, Urbana, IL 61801, USA*
[2]*Beckman Institute, University of Illinois Urbana–Champaign, Urbana, IL 61801, USA*
[3]*Department of Aerospace Engineering, University of Illinois Urbana–Champaign, Urbana, IL 61801, USA*
[4]*Department of Materials Science and Engineering, University of Illinois Urbana–Champaign, Urbana, IL 61801, USA*

9.1 INTRODUCTION

Damage in polymeric coatings, adhesives, microelectronic components and structural composites can span many length scales. Some common examples of crack damage and the corresponding length scales are summarized in Figure 9.1. Structural composites subject to impact loading can sustain significant damage on the order of tens of centimeters, which in turn can lead to subsurface millimeter-scale delaminations and micron-scale matrix cracking

Adaptive Structures: Engineering Applications   Edited by D. Wagg, I. Bond, P. Weaver and M. Friswell
© 2007 John Wiley & Sons, Ltd

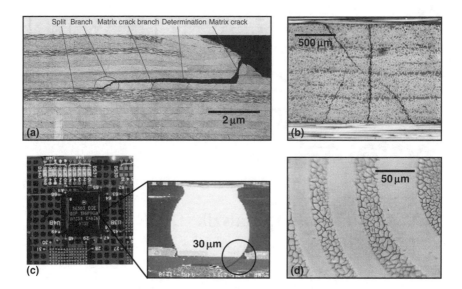

Figure 9.1 Size scales for different types of damage in polymers and polymer composites: (a) impact delamination in a graphite epoxy composite (Cvitkovich *et al.*, 1998 used with permission); (b) trans-ply cracking in a graphite epoxy composites (Kim *et al.*, 2000 used with permission); (c) compression-induced failure of an epoxy dielectric layer under a solder ball in a flip chip; (d) cracking in a patterned 50 nm thick polymer coating (Mikalsen, A.E, 1996)

(Figure 9.1(a)). Coatings and microelectronic packaging components have cracks that initiate on even smaller scales (Figure 9.1(b–c)). Repair of large-scale damage (e.g., a projectile or blast impact) is difficult and when possible requires use of bonded composite patches over the effective area. For smaller scale crack damage, however, a novel method of autonomic repair has been achieved through the use of self-healing polymers (White *et al.*, 2001).

9.1.1 Microcapsule-Based Self-Healing

Figure 9.2 illustrates the autonomic healing concept developed at the University of Illinois (White *et al.*, 2001). Crack healing is accomplished by dispersing capsules containing a healing agent and a solid catalyst within a polymer matrix. Damage in the form of a crack serves as the triggering mechanism for self-healing as does the fracture event in biological systems. The approaching crack ruptures the embedded microcapsules, releasing healing agent into the crack plane through capillary action. Polymerization of the healing agent is initiated by contact with the embedded catalyst, bonding the crack faces.

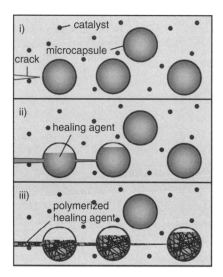

Figure 9.2   Schematic of self-healing concept (White *et al.* 2001)

Self-healing polymers and composites that incorporate microencapsulated healing agents have exhibited high levels of healing efficiency in both static and dynamic loading conditions (White *et al.*, 2001, Brown *et al.*, 2002, Brown *et al.*, 2004, Brown *et al.*, 2005a, Brown *et al.*, 2005b, Kessler and White, 2001, Kessler *et al.*, 2003). The efficiency of crack healing is defined based on the ability of a healed sample to recover fracture toughness (K_{Ic}). Fracture toughness is measured using a tapered double-cantilever beam (TDCB) test, which ensures controlled crack growth along the centerline of the brittle epoxy specimen (see inset Figure 9.3). The tapered geometry,

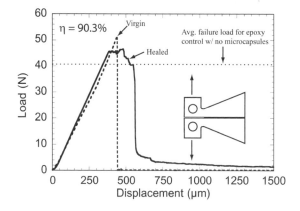

Figure 9.3   Load–displacement data for a virgin and healed epoxy fracture toughness specimen (White *et al.*, 2002 used with permission)

first introduced by Mostovoy *et al.* (1967), is designed so that the sample compliance changes linearly with crack length and the fracture toughness measurement is independent of crack length. The healing efficiency η for this geometry is expressed simply as a ratio of the peak loads (P_c) of the healed and virgin samples,

$$\eta = \frac{K_{Ic_{healed}}}{K_{Ic_{virgin}}} = \frac{P_{c_{healed}}}{P_{c_{virgin}}} \qquad (9.1)$$

Conclusive demonstration of self-healing in an epoxy resin was first obtained with a healing agent based on the ring-opening metathesis polymerization (ROMP) reaction (White *et al.*, 2001, Brown *et al.*, 2002). Dicyclopentadiene (DCPD), a highly stable monomer with excellent shelf life, was encapsulated in 180 μm diameter urea formaldehyde microcapsules with a 200 nm thick shell wall. Fracture test results using the ROMP-based healing agent are shown in Figure 9.3. For these tests, 2.5 wt % of Grubbs' catalyst was combined with EPON 828 epoxy resin (diglycidyl ether of bisphenol-A) and DETA (diethyl triamine) curing agent. To this matrix was added 5 wt % of microcapsules containing DCPD monomer. The initial (virgin) fracture toughness was assessed by propagating a pre-crack along the centerline of the specimen. The embedded microcapsules were shown to rupture in the presence of a crack and release the DCPD monomer into the crack plane. After the load was removed, the crack closed and contact with the embedded Grubbs' catalyst initiated polymerization of the DCPD. After 10 hours, the test was repeated and a healed fracture toughness was obtained. Over 90 % of the critical load was recovered in the healed samples. The dashed horizontal line in Fig. 9.3 shows the average peak load of epoxy control samples with no self-healing capability (no microcapsules). Comparison of the peak load for the control with that of the self-healing epoxy reveals that the addition of the microcapsules significantly toughens the epoxy. The self-healed samples actually recover more than 100 % of the critical load for virgin epoxy with no microcapsules.

9.1.2 Critical Issues for Microencapsulated Healing

Effective healing is dependent on three primary factors: the type of damage a material sustains, the scale of this damage, and healing kinetics. The first factor generally manifests as a distinction between brittle and ductile failure. Brittle fracture has many features which promote a quality self-healing response: the fracture surfaces are generally geometrical mirror images and allow for good contact between the two failed surfaces. Ductile failure tends to present serious challenges for self-healing materials. The damage zone is typically larger than in a comparable brittle system and the failure surfaces are not always simple to re-register, producing incomplete contact and subsequently poor bonding, which impacts self-healing performance. Materials that draw are a good example of damage that a microcapsule-based self-healing material would have difficulty effectively healing.

Damage scale, closely related to failure mode, is another critical issue for microcapsule-based self-healing materials. For a self-healing material, the damage scale is gauged primarily by the spacing between the two failure surfaces. If the spacing is larger than the critical gap that can be bridged by the volume of released healing agent, then the repairing adhesive layer can be incomplete. At the other extreme, very small cracks can fail to rupture large microcapsules, thus preventing healing agent from being released. A good understanding of the type of damage to be healed is critical when determining optimal microcapsule diameter. For a material where the self-healing mechanism is attempting to arrest and repair microcracking, a large concentration of small, possibly submicron, capsules are needed. If the damage is expected to have large failure surface separations, such as delaminations, larger microcapsules, perhaps on the order of hundreds of microns, will be required. The final factor, healing kinetics, will be discussed below.

9.2   FASTER HEALING SYSTEMS – FATIGUE LOADING

Kinetics, especially the interplay between chemical kinetics of healing and mechanical kinetics of damage, can factor heavily in the overall performance of a self-healing material. In general, faster kinetics are better; however, there are several limiting factors which determine the ideal kinetics for a given self-healing system. For heterogeneous healing chemistries, such as Grubbs' catalyst, DCPD system, the reaction kinetics must be balanced against the catalyst dissolution rate (Jones *et al.*, 2006). If the reaction proceeds too quickly, the catalyst is encapsulated by polymerized healing fluid and can no longer initiate polymerization, seriously impacting the efficiency of the system. Healing kinetics are, however, optimized when the chemical kinetics are either faster than or similar to the mechanical kinetics of damage. This effect is most pronounced in dynamic systems, such as the healing of fatigue propagated cracks.

Fatigue damage has been successfully addressed by the incorporation of a microcapsule-based self-healing system (Brown *et al.*, 2005b, Brown *et al.*, 2005a). In addition to being an important application of self-healing, fatigue damage is particularly suited to studying the interplay between the mechanical damage kinetics and the chemical kinetics of the healing chemistry. To study this balance, TDCB samples were tested in tensile fatigue using three different levels of applied range of stress intensity ΔK_I by varying K_{max} (Jones *et al.*, 2007). At high loading levels (Figure 9.4(a)), the mechanical kinetics dominate, and sufficient periods of rest must be incorporated into the training cycle for the materials system to adapt. At lower loading levels

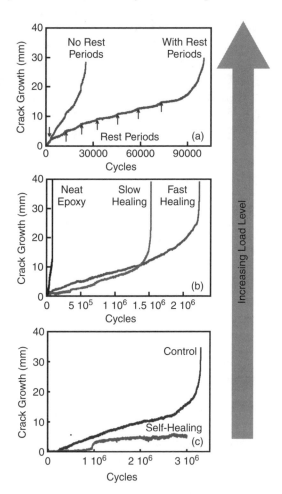

Figure 9.4 Fatigue life extension in self-healing polymers. (a) Case of high stress intensity ($\Delta K = 0.72\,\text{MPa}\,\sqrt{\text{m}}$) where mechanical kinetics dominate and rapid crack growth occurs. By incorporating periodic rest periods (training) the crack is retarded and fatigue life is extended. (b) Case of intermediate applied stress intensity ($\Delta K = 0.61\,\text{MPa}\,\sqrt{\text{m}}$) where life extension is greater for faster healing kinetics, in great contrast to a neat epoxy sample. (c) Case of relatively low stress intensity ($\Delta K = 0.50\,\text{MPa}\,\sqrt{\text{m}}$) where chemical (healing) kinetics dominate and slow crack growth occurs. In the self-healing specimen crack growth is arrested (Jones *et al.*, 2007)

where chemical kinetics dominate, we show that even under continuous fatigue the system autonomically responds by increasing its endurance limit and permanently arresting crack growth (Figure 9.4(c)). At intermediate loading levels, self-healing systems that are chemically tuned to have faster healing kinetics provide greater life extension (Figure 9.4(b)). In general, if the healing kinetics are sufficiently rapid, fatigue cracks are effectively arrested as the material is loaded.

9.3   SMALLER SIZE SCALES

While the demonstration of healing relatively large cracks, such as under monotonic fracture and fatigue loading described above, has been quite successful, self-healing may find best application when healing micron and nanometer fracture damage. Figure 9.5 shows an example of nanoscale cracking located below a fracture plane. Healing at this scale will be most effective using a capsule–catalyst system with a size scale on the order of a few nanometers. Nanometer-scaled healing system components also promise to allow for better system integration into a wide variety of materials. Fiber-reinforced composite materials, for example, will benefit by better dispersal of healing components. Submicron capsules will be able to reside in the matrix zones between individual fibers, with the aim of allowing for repair of the microcracking that frequently initiates near fibers. Thin polymeric films, such as those in microelectronics packaging, require microcapsules on these scales so the system will fit in the limited material thickness.

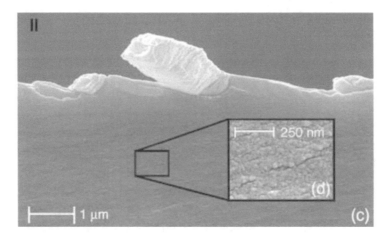

Figure 9.5   Image of TDCB fracture surface with an inset image (d) of nanoscale cracking below fracture surface (Brown *et al.*, 2004 used with permission)

A key issue for achieving healing on the nanoscale is the manufacture of nanocapsules or vessels to store the healing agents and/or catalyst. Figure 9.6 provides a schematic roadmap of encapsulation techniques and proposed strategies. Current methods of encapsulation via a macroemulsion interfacial polymerization can produce capsules ranging from 1000 μm down to a limit of 10 μm. The use of ultrasonication combined with miniemulsion techniques has significant potential to produce micron and submicron capsules as small as 800 nm in diameter. However, the development of new approaches such as colloidal templating is essential for achieving sub 200 nm nanocapsules or vessels. In addition to smaller capsules, the transition to healing on the nanoscale presents several technical challenges that include (i) selection of an optimal healing chemistry for this scale, (ii) controlled dispersion/placement of the nanocapsules in the host material, (iii) characterization of the effects of capsules on mechanical performance, and (iv) assessment of the rupture, release and healing mechanisms at this scale.

Extending self-healing to the nanoscale creates opportunities and challenges with respect to the healing chemistry. Successful healing has been achieved on the microscale via a ROMP reaction of encapsulated endo-DCPD monomer and first-generation Grubbs' catalyst. The DCPD monomer is encapsulated in a brittle urea–formaldehyde (UF) shell, and the catalyst is encapsulated in a protective wax. Both components are 100–200 μm in diameter and are easily dispersed in the matrix. Although this system meets many of the characteristics desired in a healing system, we recognize that it has limitations that must be addressed in order for self-healing to be achieved on the nanoscale.

There are several promising processing routes for making submicron DCPD capsules and catalyst beads (Rule *et al.*, 2005). However, as the size of the Grubbs' catalyst particles decreases, the reactivity also decreases significantly. Deactivation of the smaller catalyst particles is attributed to

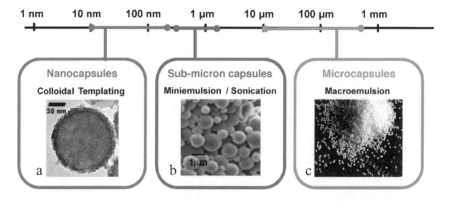

Figure 9.6 Nanotechnology roadmap of encapsulation strategies for different length scales (image a from Woloosimk *et al.*, 2005 used with permission)

increased surface area of the catalyst being exposed to unreacted amines during the cure process of the host epoxy matrix. In addition, the kinetics of the ROMP reaction have been kept fairly slow so that the healing agent is able to cover the entire plane of crack damage (about $1000\,mm^2$). For smaller scale damage, the speed of healing can be increased significantly since the healing area will be significantly reduced. While advances with ROMP-based healing agents can be anticipated, it is unlikely that ROMP chemistry will fully meet all the limitations. We thus recognize the need to broaden the concept to other chemistries to achieve self-healing on the nanoscale.

UF microcapsules prepared by *in-situ* polymerization in an oil-in-water macroemulsion have proven successful in self-healing epoxy. The diameter of the microcapsules depends on the size of the immiscible droplets in the emulsion, which can be adjusted by changing the shear rate of mixing. Capsule diameter decreases with stir speed in the form of a power law. However, for standard macroemulsion encapsulation, a lower limit of size scale is reached at $10\,\mu m$ due to limitations of stir speed.

Submicron capsules have been achieved through the use of miniemulsion techniques developed for drug delivery, food science and other encapsulation applications (Lansalot *et al.*, 2002, Asua, 2002, Tiarks *et al.*, 2001, Schork *et al.*, 2005). Miniemulsions are produced by the combination of a high shear device to break up a coarse emulsion into submicron monomer droplets with a surfactant/co-stabilizer to retard monomer diffusion from the submicron monomer droplets. We have explored the application of a sonication miniemulsion technique to produce DCPD-filled UF capsules with $1.6\,\mu m$ average diameter (Blaiszik *et al.*, 2007) – an order of magnitude reduction in capsule diameter (Figure 9.6). These results are highly promising given that this reduction in capsule diameter was achieved without a co-stabilizer. Also of importance is the rupture of the capsules and subsequent release of healing agent at these size scales. SEM images of $1.6\,\mu m$ diameter capsules in a fracture plane verify rupture at this smaller diameter (Figure 9.7). Also promising is the appearance of 'tails' on the fracture surface, indicating that a crack-pinning toughening mechanism is operative as well.

9.4 ALTERNATIVE MATERIALS SYSTEMS – ELASTOMERS

To date, the focus of self-healing materials has been on the incorporation of an autonomic repair functionality into brittle, thermoset materials. In general, these materials, as discussed above, have several characteristics which make them good candidates for the incorporation of self-healing functionalities. While differing significantly in elastic performance, elastomeric materials also possess several characteristics which allow for efficient self-healing.

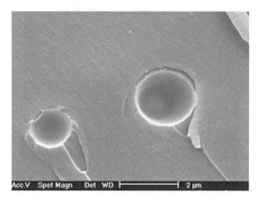

Figure 9.7   SEM image of rupture capsules in an epoxy matrix

Elastomers do not fail in a brittle manner, but tear, failures which share some characteristics with brittle fracture. Elastomeric failures generally do not involve large, permanent deformations and produce easily re-registerable failure surfaces. Many elastomers are also processed in a liquid state which allows for the incorporation of microcapsules prior to chemical crosslinking.

Self-healing in elastomeric materials has been demonstrated through the incorporation of a microcapsule-based self-healing system in a poly(dimethylsiloxane) (PDMS) elastomer (Keller *et al.*, 2007). The self-healing PDMS contains two microcapsule systems, a resin microcapsule containing a high-molecular-weight reactive PDMS and a low-concentration catalyst and an initiator capsule, which contains a crosslinking co-polymer. Both materials were encapsulated in a UF shell, as in the epoxy system discussed above. Triggering of the self-healing function proceeds similarly to the system described above, but instead of a heterogeneous solid–liquid interaction, the two fluids, resin and initiator, interact on the crack plane to produce the polymer which bonds the failure surfaces together.

Testing and verification of self-healing PDMS was performed using a trouser tear specimen and quantified using a criterion similar to the one described above in Equation (9.1) for thermosetting materials. For elastomer healing the healing efficiency is defined in terms of the ratio between the virgin and healed tear strengths

$$\eta = \frac{T_{healed}}{T_{virgin}} = \frac{F_{avg_{healed}}}{F_{avg_{virgin}}} \tag{9.2}$$

which collapses to the ratios of the average tearing force.

The system described above has shown the capability to effectively repair tear damage produced by a trouser tear specimen and can, in certain cases, completely heal the material back to the original strength. Figure 9.8 shows

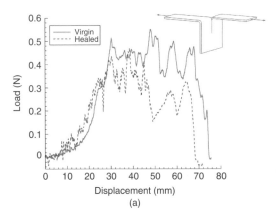

(a)

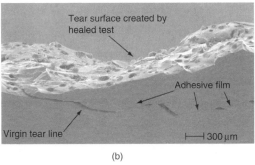

(b)

Figure 9.8 (a) Load–displacement data for a virgin and healed PDMS tear specimen. Healing efficiency is 76 % in this case. (b) Side view of a healed tear specimen showing the adhesive zones produced by healing fluid released from ruptured microcapsules (Keller *et al.*, 2007)

representative load–displacement data for a virgin and healed test of the self-healing elastomeric material. Current work in this area is focusing on the fatigue response of this self-healing material and incorporation into bladder structures for space and terrestrial applications.

9.5   MICROVASCULAR AUTONOMIC COMPOSITES

While the current generation of microencapsulated self-healing polymers and composites has proven remarkably successful, some limitations are inherent to this conceptual approach. Eventually the healing functionality will be compromised as the supply of healing agent is consumed. Natural systems overcome this limitation by incorporating a pervasive vascular network of pathways (a circulatory network) to supply the chemical components for healing on demand. Of course, in nature this circulatory network satisfies many other functions as well (e.g., thermal regulation, growth, chemical sensing). Future generations of autonomic materials will undoubtedly incorporate these principles to achieve unparalleled performance and multifunctionality.

In moving from a compartmental (microcapsules) to a circulatory approach the primary technical challenge is to design and fabricate a pervasive, inter-connected, three-dimensional (3-D) vascular network across multiple length scales. In natural systems, extremely complex 3-D vascular tree structures have evolved (Figure 9.9). While this level of complexity is beyond current engineering practice, simplified designs have recently been fabricated and tested (Toohey *et al.*, 2007). Advancements in microfabrication techniques and materials integration will enable more sophisticated structural materials in the near future.

With circulatory networks incorporated in structural materials systems a variety of functionality is enabled. For self-healing materials the circulation of healing agent throughout the vasculature provides large volumes of repair

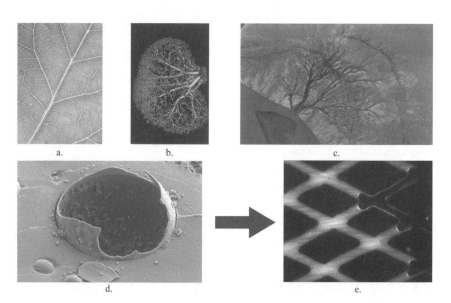

Figure 9.9 Natural vascular structures and the future development of synthetic autonomic materials. Natural systems include the (a) leaf, (b) rabbit kidney (Seruetz *et al.*, 1995 used with permission), and (c) drainage basin (U.S. Navy photo). The first generation of autonomic healing is based on compartmentalization (via micro-capsules) of the healing agent as shown in (d). In the future the healing agent will be supplied via a microvascular network as shown in this fluorescent microscope image of a dye moving through a two-dimensional network embedded in structural epoxy (e) (Therriault, 2003)

chemicals delivered on demand to the site of damage. If the network can be replenished periodically then the supply of healing agent is unlimited in practice. For thermal management the circulation of a cooling/heating fluid or gas phase can efficiently remove or deliver heat to critical areas. Sensory molecules can be incorporated in the fluid stream to act as chemical or environmental sensors as well. Ultimately, a single circulatory network could provide all of these functions in a single vascular system.

9.6 CONCLUSIONS

Microcapsule-based self-healing materials have provided the first demonstration of fully autonomic material repair. The concept, assuming a suitable microcapsule system can be found, is general and has been applied to new material classes, specifically elastomers. Advances in capsule manufacture have allowed for production of micron and submicron capsules, which will enable new applications of self-healing, especially in such areas as fiber-reinforced composites and thin films. While microcapsule-based self-healing has shown that autonomic repair of a material is possible, future work will increasingly focus on the incorporation of microvascular networks as the basis of a self-healing material. These microvascular networks will allow for continuous regeneration of the healing fluids and will allow for other functionalities, such as cooling or sensing, to be incorporated into a material.

References

ASUA, J.M. (2002) Miniemulsion polymerization. *Progress in Polymer Science*, 27, 1283–1346.

BLAISZIK, B., WHITE, S.R. & SOTTOS, N.R. (2007) Nanocapsules for self-healing composites. Submitted to *Composites Science and Technology*.

BROWN, E.N., SOTTOS, N.R. & WHITE, S.R. (2002) Fracture testing of a self-healing polymer composite. *Experimental Mechanics*, 42, 372–379.

BROWN, E.N., WHITE, S.R. & SOTTOS, N.R. (2004) Microcapsule induced toughening in a self-healing polymer composite. *Journal of Materials Science*, 39, 1703–1710.

BROWN, E.N., WHITE, S.R. & SOTTOS, N.R. (2005a) Retardation and repair of fatigue cracks in a microcapsule toughened epoxy composite - Part 1: Manual infiltration. *Composites Science and Technology*, 65, 2466–2473.

BROWN, E.N., WHITE, S.R. & SOTTOS, N.R. (2005b) Retardation and repair of fatigue cracks in a microcapsule toughened epoxy composite - Part II: In situ self-healing. *Composites Science and Technology*, 65, 2474–2480.

CVITKOVICH, M. K., O' BRIEN, T.K. & MINGUET, P.J. (1998) Fatigue delouding characterization in composite skin/stringer configurature. In *composite Materials: Fatigue and fracture*, 7, 97-121.

JONES, A.S., WHITE, S.R., RULE, J.D. & SOTTOS, N.R. (2007) Influence of chemical kinetics on fatigue life extension in a self-healing polymer. *Journal of the Royal Society Interface* (on-line), 4, 395–403.

JONES, A.S., WHITE, S.R., SOTTOS, N.R., RULE, J.D. & MOORE, J.S. (2006) Catalyst morphology and dissolution kinetics for self-healing polymers. *Chemistry of Materials*, 18, 1312–1317.

KELLER, M.W., WHITE, S.R. & SOTTOS, N.R. (2007) A self-healing Poly(dimethylsiloxane) eilastomer. Accepted to *Advanced Functional Materials*.

KESSLER, M.R., SOTTOS, N.R. & WHITE, S.R. (2003) Self-healing structural composite materials. *Composites Part A-Applied Science and Manufacturing*, 34, 743–753.

KESSLER, M.R. & WHITE, S.R. (2001) Self-activated healing of delamination damage in woven composites. *Composites Part A-Applied Science and Manufacturing*, 32, 683–699.

KIM, R.Y., CRASTO, A.S. & SCHOEPPNER, G.A. (2000) Dimensional stability of composite in a space thermal environment. *Composites Science and Technology*, 60, 2601–2608.

LANSALOT, M., DAVIS, T.P. & HEUTS, J.P. A. (2002) RAFT miniemulsion polymerization: Influence of the structure of the RAFT agent. *Macromolecules*, 35, 7482–7591.

MOSTOVOY, S., CROSLEY, P.B. & RIPLING, E.J. (1967) Use of crack-line loaded specimens for measuring plane strain fracture toughness. *Journal of Materials*, 2, 661–681.

MIKALSEN, A.E. (1996) Mediated patterning of sol-gel thin layers: shrinkage, de-cohesion, and lift-off, PhD dissertation. University of Illinois at Urbana-Champaign.

RULE, J.D., BROWN, E.N., SOTTOS, N.R., WHITE, S.R. & MOORE, J.S. (2005) Wax-protected catalyst microspheres for efficient self-healing materials. *Advanced Materials*, 17, 205–208.

SCHORK, F.J., LUO, Y.W., SMULDERS, W., RUSSUM, J.P., BUTTE, A. & FONTENOT, K. (2005) Miniemulsion polymerization. *Polymer Particles*, Advances in Polymer Science 175, 129–255, Springer.

SERNETZ, M., JUSTEN, M. & JESTCZEMSKI, F. (1995) Dispersive characterization of kidney arteries by three dimensional mass-radius-analysis. *Fractals*, 3, 879–891.

THERRIAULT, D., (2003) Directed assembly of three-dimensional microvascular networks, PhD dissertation. University of Illinois at Urbana-Champaign.

TIARKS, F., LANDFESTER, K. & ANTONIETTI, M. (2001) Preparation of nanocapsules by miniemulsion polymerization. *Langmuir*, 17, 908–918.

TOOHEY, K.S., SOTTOS, N.R., WHITE, S.R., LEWIS, J.A. & MOORE, J.S. (2007). Self-healing materials with microvascular networks, in review for Nature Materials.

WHITE, S.R., SOTTOS, N.R., GEUBELLE, P.H., MOORE, J.S., KESSLER, M.R., SRIRAM, S.R., BROWN, E.N. & VISWANATHAN, S. (2001) Autonomic healing of polymer composites. *Nature*, 409, 794–797.

WOLOSIUK, A., ARMAGAN, O., & BRAUN, P.V. (2005) Double direct templating of periodically nanostructured ZnS hollow microspheres. *Journal of the American Chemical Society*, 127, 16356–16357.

10

Adaptive Structures – Some Biological Paradigms

Julian F.V. Vincent

Department of Mechanical Engineering, The University, Bath BA2 7AY, UK

10.1 INTRODUCTION

Ideas have been interchanged between biology and engineering for a long time. When the traffic is from engineering to biology it is commonly known as biomechanics; in the other direction it is known as biomimetics or biomimicry or bionics or bioinspiration. But probably because the manipulation of the world around us is always done from the perspective of an engineer, the rules of this interchange are always laid down by the engineer. Even when an idea is taken from nature and made to work for us in our technology, the end point is more recognisable as engineering than as biology. Thus the hooks in Velcro are used for joining textiles, not for getting a free ride which is what the hooked seed, which suggested Velcro, is doing; the growth shapes of trees are converted into designs for bridges and cars and are not used to guide us into growing durable composite structures. But why do we wish to steal concepts from biology anyway? It is because the 'design' which has evolved is good (in terms of *engineering* design) and because often the mechanisms used are either novel or being used in

Adaptive Structures: Engineering Applications   Edited by D. Wagg, I. Bond, P. Weaver and M. Friswell
© 2007 John Wiley & Sons, Ltd

a novel way – at least by our standards. Increasingly I consider this not to be the best use of biology in engineering. I want to know not just how a function or mechanism can be implemented in the world of the engineer, but how biology would solve the problem which the engineer is trying to solve. However, the focused methods used in the engineering world prove to be an interesting comparison with the rather broader way in which biological organisms produce similar effects. Thus a GFRP plate showing bistable curvature is produced in only one main way. Biology has several ways of producing the same effect, in both living and dead tissues and organs. This complexity is commonplace to biologists, and the engineering approach is useful in that its rigour serves to distil some order out of the apparent chaos of biology. However, the biological systems in general do two things that engineering tends not to. First, all biological functions have to evolve from pre-existing conditions. Thus any function (such as adaptive morphological change) will be achieved in a number of different ways depending on the ancestor's adaptation, phylogenetic freedom, biochemical and physiological mechanisms, etc. This variety may suggest useful alternatives to an engineer faced with specific design problems. This is illustrated here by the discussion of how plants move. Second, the tendency is for biological mechanisms to be only just good enough (taking into account the familiar optimisations of expense, safety, repair, etc.), so quite often some intriguingly simple solutions appear. This point is illustrated by the strain sensors found in arthropods.

Technology is leading us down paths which are potentially dangerous for humanity, one of which is global warming. Biological systems are, by evolution and definition, sustainably adapted to life on a maintainable Earth. Comparing biological and technical methods it is clear that there is much similarity between biological systems and adaptive engineering structures. My final point in this essay, then, is that we must bend our engineering to be more biological – to manipulate our world from the perspective of the biologist and not the engineer. Adaptive structures look to be a very good way of furthering this aim.

10.2   DEPLOYMENT

Although many folding patterns can be found in plant structures (Kobayashi *et al.*, 1998; Kresling, 1991; Kresling, 2000) there have been few mechanical studies. The leaves of many plants, especially broad-leaved trees of temperate areas, are folded or rolled while inside the bud. For example, the leaves of hornbeam and beech have a straight central vein and symmetrically arranged parallel lateral veins which generate a corrugated surface. The central vein elongates, separating the bases of the lateral veins and causing the lamina between the lateral veins to rotate into the plane of the leaf, at the same time causing the lateral veins to rotate away from the central vein (Figure 10.1). These two mechanisms provide the initial fifth of the increase in the projected area of the leaf as it expands. Other leaves such as sycamore and maple unfold in a more radial manner. In the leaf, the membrane between the veins also expands. The controlling factor here seems to be the orientation of the cellulose microfibrils in the walls of the cells which make the upper and lower surfaces of the leaf (the epidermal cells). In the early expansion phase the cellulose is orientated orthogonally to the direction in which expansion will occur, so that only the material between the cellulose fibres, of lower modulus, needs to the stretched. When expansion finishes, the cellulose fibres have rotated 90 degrees, so stiffening the membrane in the expansion direction and stopping the process.

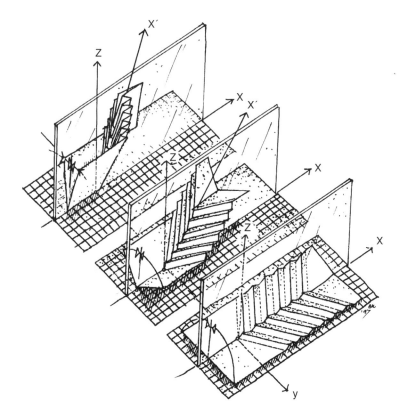

Figure 10.1 A paper-folding ('ha-ori') of the main mechanism of opening of beech and hornbeam leaves (Kobayashi *et al.*, 1998)

The unfolding of flower petals employs a number of mechanisms not found in leaves. A Miura-ori (Miura, 1993) is evident in poppy petals (Delarue, 1991); the flowers of hollyhock and morning glory use a spiral packing mechanism (Guest and Pellegrino, 1992).

The geometry and mechanics of wing folding of beetles have been studied using vector analysis (Haas and Beutel, 2001). Although in general the patterns of folding follow simple rules, it is often important for the wing to be mechanically bistable, since its folding can be controlled only by three hinge points at the base of the wing. It is therefore quite common to find buckling mechanisms built into the wing structure which can both stiffen the membrane and turn it into a bistable mechanism, enabling it to fold and unfold and to remain in either of those states when required. These mechanisms have been identified in general terms, but not analysed. At least in part this is because the mechanical properties of the wing membrane and stiffening structures, collectively made of a composite of chitin nanofibres in a tanned protein matrix, are complex (Herbert *et al.*, 2000; Smith *et al.*, 2000; Wootton *et al.*, 2000).

10.3 TURGOR-DRIVEN MECHANISMS

The cells of all living plants maintain shape and size using internal or turgor pressure, especially at early stages of their development. If a developed cell does not lignify fully, it needs to supplement its rigidity with turgor. Since plants are essentially hydraulic machines which can generate quite surprising forces from sugar solutions (Beukers and Hinte, 1998), it is worth examining their pressure system. A simple model of the turgor system starts from observations of the hollow flowering stem of the dandelion, *Taraxacum officinale* (Vincent and Jeronimidis, 1992), which is made, as are all multicellular plants, of cells confined within stiff cell walls composed mainly of cellulose ($E = 130\,\text{GPa}$). The distribution of cellulose across the stem is graded. On the outside of the stem the cells are small (radius $r = 3\,\mu\text{m}$) and the thickness of the cell walls, t, is about $3\,\mu\text{m}$; on the inside of the stem the cells are larger ($r = 30\,\mu\text{m}$) with t less than $1\,\mu\text{m}$. Thus the volume fraction of cell wall material varies from about 0.5 on the outside of the stem to about 0.013 in the inside with an average of 0.0844. This corresponds to a mean mass fraction of cellulose of 8.8 % assuming that the cell walls contain 30 % of water in the cellulose, which itself has a dry density of $1500\,\text{kg/m}^3$. This compares with a measured dry matter of 8.73 %. Strips cut longitudinally from the stem curl by differing amounts when immersed in solutions of sugar varying from 0 to 15 %; in a totally wilted stem the epidermis is crinkled, whereas in the fully turgid stem it is smooth. These observations can be used as the basis of a model, making the following assumptions:

(a) all cells are at the same turgor and have the same osmotic potential;
(b) all cell walls have the same stiffness, which is proportional to the amount of cellulose in them;
(c) all cells are cylindrical along the length of the stem;
(d) the cells are firmly stuck together.

The following model (Vincent and Jeronimidis, 1992) is due mainly to Jeronimidis. We establish a positional reference, z, which refers to the mid-plane of the section of the stem ($z = 0$) with the positive direction towards the epidermis. The stem is notionally divided into a large number of strips so that within each strip, identified by its position in the stem ($+z$ or $-z$), the volume fraction of cell wall material, V_f^{cw}, can be considered constant. The volume fraction of cell wall material at any place across the stem is then given by

$$V_f^{cw}(z) = \frac{1}{4}\pi[(2\Phi(z) - 1)/(\Phi(z)^2)] \tag{10.1}$$

where $\Phi(z) = r(z)/t(z)$, and the Young's modulus of each strip is

$$E_x(z) = V_f^{cw}(z)E_{cw} \tag{10.2}$$

where E_{cw} is the Young's modulus of the cell wall material in the axial direction. This was measured directly on a non-turgid stem as about 500 MPa. The response of this multi-layered stem to external forces and moments depends on three parameters from the theory for laminate composite beams:

$$A = \int_{-1/2h}^{+1/2h} E_x(z)\, dz \tag{10.3a}$$

which relates the axial force N_x per unit width to the mid-plane ($z = 0$) strains;

$$B = \int_{-1/2h}^{+1/2h} E_x(z)z\, dz \tag{10.3b}$$

which takes into account the fact that in a non-symmetrical beam the neutral axis of bending is not located at the centroid of the section (the mid-plane);

$$D = \int_{-1/2h}^{+1/2h} E_x(z)z^2 \, dz \qquad (10.3c)$$

which relates the bending moment per unit width, M_x, to the curvature of the mid-plane. The axial force and bending moment are related to the mid-plane strains and curvatures by

$$N_x = A\varepsilon_x^0 + BK_x \qquad (10.4a)$$

and

$$M_x = B\varepsilon_x^0 + DK_x \qquad (10.4b)$$

where ε_x^0 is the mid-plane normal strain and K_x is the mid-plane curvature. By substitution from (10.2) into (10.3a–c) and by numerical integration (which required the thickness to be divided into at least 600 strips) values for A, B and D were obtained. In the absence of external forces and moments, a longitudinal strip cut from a turgid stem has a curvature K_x, so that the effect of the turgor can be expressed in much the same way that thermal forces and moments are used in the analysis of the mechanics of a bimetallic strip. However, we still need to introduce the turgor pressure, P, into the relations. The axial strain, ε_x^{cw}, induced in the cells by free axial expansion due to turgor, assuming that radial expansion is limited both by neighbouring cells and by the presence of a stiff epidermis, is given by

$$\varepsilon_x^{cw} = \beta(P/E_{cw})(r/t) \qquad (10.5a)$$

where the coefficient β is a measure of the radial restraint. For full restraint,

$$\beta = 1/2(1 - \nu_{yx}\nu_{xy}) \qquad (10.5b)$$

If the Poisson ratio ν_{xy} in the cell wall is 0.3, then ν_{yx} will be at least 10 times smaller, their product will be small, and β can be taken as 0.5. This equation can be used to define a coefficient of pressure expansion $\alpha_x^P(z)$, analogous to a coefficient of thermal expansion, in the form

$$\alpha_x^P(z) = \varepsilon_x^{cw}/P = \beta(1/E_{cw})(r/t)$$
$$= \beta\Phi(z)/E_{cw}$$
$$= 0.5\Phi(z)/E_{cw} \qquad (10.6)$$

This relation can be used to calculate the equivalent forces and moments which produce curvature in a strip excised longitudinally from the stem under turgor pressure ΔP (with reference value $P = 0$ in the fully wilted state); thus

$$N_x^P = \int_{-1/2h}^{+1/2h} E_x(z)\alpha_x^P(z)\Delta P \; dz \qquad (10.7)$$

and

$$M_x^P = \int_{-1/2h}^{+1/2h} E_x(z)\alpha_x^P(z)z\Delta P \; dz \qquad (10.8)$$

Numerical values for N_x^P and M_x^P are obtained by introducing (10.1), (10.2) and (10.6) into (10.7) and (10.8) and then integrating. These data can be used to calculate the curvatures and mid-plane strains due to turgor pressure using (10.3a–c) and (10.4a,b) as

$$K_x^P = [(AM_x^P - BN_x^P)/(AD - B^2)]\Delta P/E_{cw} \qquad (10.9)$$

and

$$\varepsilon_x^{OP} = [(DN_x^P - BM_x^P)/(AD - B^2)]\Delta P/E_{cw} \qquad (10.10)$$

The inverse of curvature, K, is the radius of curvature which is more easily measured on excised strips of stem placed in solutions of mannitol (a sugar which is not metabolised by the plant cells) and can now be calculated if P is known. The concentration of sugar required to cause an excised strip of dandelion flowering stem to straighten has an osmotic potential of about 20 bar. This will be the turgor pressure exerted against the cell walls when the same strip is immersed in distilled water and the cells imbibe water. Calculation based on the bimetallic strip model gives a radius of curvature of 3.4 mm of a strip immersed in distilled water, compared with an experimental value of 3 mm. This gives us a quantitative model of the effect of turgor pressure on the properties of plant tissues. The reason why the excised strip curls is that the cell walls of the inner cells are much thinner than those of the outer cells. They will thus stretch further under the same turgor pressure since the stress will be greater on the inner wall. Thus the excised strip will curl with the thinner walled cells on the outside of the curl. The resultant strain at any point throughout the thickness of the intact stem is the sum of the strains due to turgor pressure and to the bending moments reflecting the constraints which hold each strip straight while it is still part of the stem. Thus

$$\{\varepsilon_x(z)\}_{total} = (BK_x^P/A + \varepsilon_x^0)\Delta P/E_{cw} \qquad (10.11)$$

This is independent of z (strain does not depend on the position across the stem), which is to be expected since the stem is straight. The net stress in the stem can now be obtained by assuming that, for small strains, the stress is proportional to the strain, a fact which can be confirmed experimentally by a tensile test of the intact wilted stem. Thus we apply Hooke's law in each layer and multiply the total strain by the local Young's modulus of the stem. The local stress will therefore vary across the stem in proportion to the local volume fraction of cellulose.

We now have a model for the generation of non-uniform pre-stress in a turgor-driven plant system. Is there anything more interesting to which we can apply it?

10.3.1 The Venus Fly Trap

The Venus Fly Trap (*Dionaea muscipula*) preys on insects and other small animals which venture onto its trap leaves and trigger their closure by disturbing certain sensitive hairs (Figure 10.2). High-speed video recording shows that the leaves routinely shut in 1/25 s. Such speed is uncommon amongst plants and so has attracted attention and theories for many years. The mechanism is based on turgor-driven elastic instability of the leaf. A better understanding of this mechanism, and the way in which it is designed and actuated, would not only solve a long-standing conundrum, but also give rise to a series of lightweight pressurised actuators, switches and morphing structures.

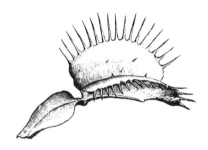

Figure 10.2 The Venus Fly Trap leaf, showing the three trigger hairs on one of the leaf halves

10.3.2   Previous Theories

It was first thought (Darwin, 1875) that the midrib of the trap is a hinge (which it is not). It was then suggested (von Guttenberg, 1959; von Guttenberg, 1971) that the upper epidermis loses turgor and shrinks, or that the cell walls of the lower epidermis soften and stretch under turgor pressure (Williams and Bennett, 1982). Eventually it was concluded that any mechanism which requires a physiological (e.g. turgor) change will be too slow (Hodick and Sievers, 1989). It now seems obvious that the only type of mechanism which can provide the speed is elastic, using stored strain energy provided by turgor pressure. This has the advantage that the turgor and mechanical properties of the cells and tissues do not need to change during the fast phase of closure, though such changes occur both in the earlier phases (Forterre *et al.*, 2005; Stuhlman, 1948) and in the later phases of closure (Fagerberg and Allain, 1991; Stuhlman, 1948). This still leaves the nature of the trigger for the instability in question, but it changes the nature of the question.

10.3.3   Background to an Elastic Model

A plate can change shape from one curvature to another (equivalent to the fly trap leaf being open or closed) without any change having taken place in its elastic properties. The shape change is elastic; often it is symmetrical, requiring the same forces and displacements to flip it from one curvature to the other. It is easy to construct a plate in which the behaviour is not symmetrical and the plate is stable in one configuration but only just stable in the other. This would be equivalent to the closed and open positions of the leaf, respectively. Thus it is possible for the fly trap leaf to go from a quasi-stable open position to a stable closed position without any change of its elastic properties, and any arguments about turgor not being fast enough are irrelevant since the global elastic properties need not change in order for the mechanism to function. High speed video shows that the shape change starts near the free edge of the leaf and travels towards the main central vein.

Simple microscopy of the leaf (Figure 10.3) shows that it has five layers: upper and lower epidermis, a layer of small cortical cells inside the epidermis, and a thick central region (medulla) of cells about 300 μm long and 40 μm diameter. Polarised light microscopy shows that these medullary cells have their cellulose microfibrils wound around them like the bands around a barrel, so that internal (turgor) pressure encourages them to elongate. Thus the leaf must be under radial pre-strain. In particular, the walls of the lower epidermis, which stretches by 7 % when the leaf closes (Hodick and Sievers,

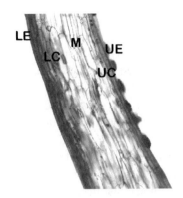

Figure 10.3   Section through a Venus Fly Trap leaf. LE, UE – Lower and Upper Epidermis; LC, UC – Lower and Upper Cortex; M – Medulla

Table 10.1

Cell layer	Thickness (μm)	V_f cellulose	% cellulose
Upper epidermis	5	0.28	2.5
Upper cortex	45	0.22	32
Medulla	365	0.07	45
Lower cortex	80	0.22	18
Lower epidermis	5	0.28	2.5

1989), are strongly pleated transversely to the long axis of the cells. These walls therefore unfold rather than stretch (7 % is a strain far greater than cellulose can endure). The thick-walled upper epidermis hardly shortens at all (Hodick and Sievers, 1989). These measurements, and the parameters derived (Table 10.1), show that there is an asymmetry of nearly 2:1 in the thickness of the cortical layer on the upper and lower surfaces.

The % cellulose is based on the volume fraction of the cell wall in the different layers, assuming them to be 70 % cellulose. This results in the mechanical parameters of Table 10.2, based on a value for the stiffness of the cell wall of 500 MPa (Probine and Preston, 1962). In the radial direction the cells will not stretch significantly, but axially the inner cells can stretch the leaf by 2 % under a turgor pressure of 10 bar. The turgor pressure of fly trap leaf cells has not been reported, but 10 bar would not be unusual, especially since this is a plant which lives in bogs with plentiful water.

As the leaf closes the curvature reduces slowly, reverses suddenly and the leaf snaps shut. If all is well, it closes around the small insect which has

Table 10.2

Cell layer	E_{11}	E_{22}	G_{12}	ν_{12}	α_{11}	α_{22}
Upper epidermis	200	200	60	0.3	0	0
Upper cortex	160	8	4	0.3	0.001	0.05
Medulla	50	2.5	1	0.3	0.005	0.2
Lower cortex	160	8	4	0.3	0.001	0.05
Lower epidermis	200	200	60	0.3	0	0

E_{11} and E_{22} are the hoop and longitudinal stiffnesses of the walls of the cells in the different layers in MPa, G_{12} is the shear modulus in MPa (estimated) and ν_{12} is the Poisson ratio in the direction in which the leaf is mainly stretched (also estimated). α_{11} and α_{22} are calculated coefficients of expansion (% strain per bar of turgor pressure) of the cells in the radial and axial directions respectively.

just been disturbing the trigger hairs. Once closed, although the leaf can be bent back into the open position, requiring a force of about 10 N, it quickly closes again, suggesting that the redistributed stresses have relaxed due to redistribution of turgor pressure or changes in the stiffness of the cell walls (Fagerberg and Allain, 1991). During this opening phase the leaf 'grows' (Fagerberg and Howe, 1996) suggesting that some cell walls have reduced in stiffness and expand under turgor.

10.3.4 The Trigger

Even with a model driven by strain energy, the prevailing view is that the trigger is provided by a change in electrical potential (Forterre *et al.*, 2005). This seems unnecessary, especially when it appears that a purely mechanical catastrophe model (Figure 10.4) is adequate to illustrate the movements. The model then becomes a study in stability, such that an elastic instability is sufficient to trigger closure. This removes the necessity for a global signalling mechanism within the entire leaf. This mechanism has frequently been proposed and searched for, but never found (Hodick and Sievers, 1989). Indeed, any model based on electrical change still requires a transduction from electrical to mechanical, and this change has to be distributed over the leaf in such a way that the shape change is uniformly graded across the leaf irrespective of the point of stimulation. This requires conduction pathways which are similarly independent of origin and which will produce a perfectly integrated response across the entire leaf. The closest analogy would be the stellate ganglion system of the squid, which causes the muscles of the mantle to contract as a unit, irrespective of their distance from the initiating impulse in the ganglion. It can do this because the nerves have

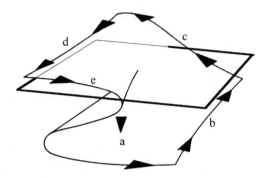

Figure 10.4 Catastrophe model (Thom, 1975) of the Venus Fly Trap leaf closing and opening cycle. Above the plane the leaf is open; below it the leaf is closed. The pathway *b* probably represents loss of turgor; the pathway *d* probably represents re-establishment of turgor. High turgor gives high pre-stress and hence stored elastic energy. The leaf closes (*a*) in a catastrophic manner but opens (*c*) smoothly

a diameter in proportion to the distance from the ganglion over which they have to transmit the signal to the muscles, a detail of morphology essential for its proper function. Any such analogous structure seems totally absent from the fly trap leaf. The simple answer has to be that no such electrical system exists and the whole functioning of the leaf is dependent on the control of its elastic stability.

10.4 DEAD PLANT TISSUES

There are many dead plant tissues which move as a result of changing humidity. Many of them appear to be laid up as composite bilayers of cellulose microfibres in orthogonally preferred orientations. An example is the pod of many legumes (e.g. gorse), which use anisotropic fibre orientations in the pod wall to build up strain energy as they dry out, with explosive brittle fracture releasing the energy with sufficient power to project the seeds. The seed pod dries to a helix. As far as I know these beautiful structures have never been analysed mechanically, although we now have the tools to do so. Another structure which alternates between straight and helical is the awn of the seed of the storksbill (Figure 10.5). When wet it is straight, but it curls when dry. If the tip of the awn snags against a soil particle, then the seed can be drilled into the ground! Many lower plants such as mosses and liverworts have spore cases which dry and split, broadcasting the spores to significant distances.

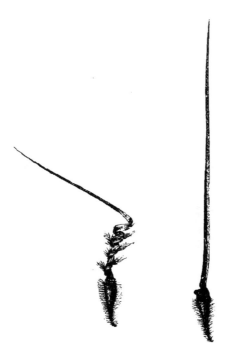

Figure 10.5   The seed of the storksbill: dry (left) and wet (right)

Another example is the pine cone whose bracts (the seed-bearing scales) open in dry weather to expose the seed, but close in damp weather (Figure 10.6) and so stop the seeds getting too moist before they are shed (Dawson *et al.*, 1997). Dissection of the scale revealed that it is composed of two types of sclerenchyma (highly lignified fibre cells) extending along the scale from the centre of the cone to the tip of the bract. The outside cells on the lower surface of the deployed bract are sclerids; on the inner or upper surface are fibres. The materials of the two layers are identical in chemistry and in water adsorption characteristics, but differ in morphology: the angle of winding of the cellulose microfibrils in the walls of the fibre cells, relative to the long axis of the cell, is much greater in the sclerids ($74° \pm 5°$) than the fibres ($30° \pm 2°$). Thus the fibres are longitudinally stiffer (4.53 ± 0.90 GPa compared with 0.86 ± 0.05 GPa). This difference in stiffness means that with a change in relative humidity of only 1 % the coefficient of hygroscopic expansion of the fibres (0.06 ± 0.02) is significantly lower than that of the sclerids (0.20 ± 0.04). Once again we have a system which can be modelled using the mathematics of the bimetallic strip, but this time the differences are not due to the volume fraction of cellulose. The mechanism of hygroscopic bending depends on the orientation of the cellulose microfibrils in the walls

Figure 10.6   A pine cone: dry (left) and wet (right)

of fibres – a hierarchical approach to organising anisotropy. There has been no further analysis of the pine cone because there are no experimental data on the coefficient of hygroscopic expansion of cellulose, lignin, hemicellulose or pectins.

10.5   MORPHING AND ADAPTING IN ANIMALS

Plants in general are passive, animals active. Whilst a discussion of movement in animals would not be productive in this context, an account of morphological change is appropriate. Embryology, development, metamorphosis are all interesting but can have little relevance; however, there is one group of animals which can change their skeletal morphology – the echinoderms. These animals (sea urchins, brittle-stars, starfish, sea cucumbers and sea lilies) can, sometimes very quickly, change the properties of the connective tissue between the hard parts of the skeleton. The equivalent would be a composite with a thermoplastic matrix which could be softened and reshaped, locally or globally, at ambient temperature. With this trick, the sea cucumber can turn itself into a viscous liquid which can be poured from one container to another, and the Crown-of-thorns Starfish can terrorise the clams of the Great Barrier Reef by draping itself over their shells, adapting to their shape, then developing sufficient force to prise open their valves and insert its stomach into their body cavity and digest them *in situ*.

10.6   SENSING IN ARTHROPODS – CAMPANIFORM AND SLIT SENSILLA

The ability of a structure to sense loads – endogenous or exogenous – is an important part of modern design and health monitoring. The way displacement detectors work in most of biology is essentially unknown, since they are small, soft and probably rely on changes in membrane permeability for their function. However, there is a simple structure in arthropods (animals with jointed limbs and an external skeleton, such as insects, spiders, scorpions, crabs, etc.) which consists of a hole through the skeleton which thus makes the immediate area more compliant and acts as a strain amplifier (Figure 10.7). In a spider, the hole is a slit or slot, about 1 μm across. Any shape change is rotated through 90° out of the plane of the cuticle by a trough-shaped covering suspended in the slit whose deflection is detected by a cell beneath the cuticle. At the cellular level, the means of detection of displacement is still speculative. Slits can occur in isolation on the surface of the spider, or in groups of a range of lengths arranged more or less parallel to each other. These often appear like the strings on a harp or lyre, hence their

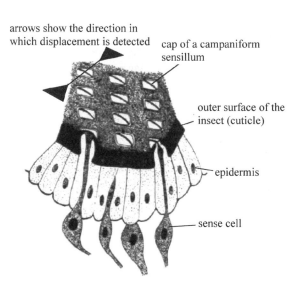

arrows show the direction in which displacement is detected

cap of a campaniform sensillum

outer surface of the insect (cuticle)

epidermis

sense cell

Figure 10.7   A section of insect cuticle showing a small field of campaniform sensilla. Each sensillum is about 5 μm across

name of 'lyriform organs'. In particular the lyriform organ is a frequency analyser which can be tuned (Baurecht and Barth, 1992). This suggests that a biomimetic slit sensor would be able to detect particular frequencies, and generate frequency information about an input signal with no further analysis (e.g. by computer).

Insects have a similar organ – the campaniform sensillum – which is a round or oval hole rather than a slit through the cuticle. Finite elements modelling of a campaniform sensillum (Skordos *et al.*, 2002) showed that global deformation of the plate (which can be flat, curved or a tube) induces higher deformation of the hole due to its higher local compliance. This local amplification of deformation offers a sensitive mechanism for sensing complex, time-varying strains, for example arising from vibrations. Further, since the local deformation of a sensillum is frequency dependent there is a real prospect of determining the spectral distribution of energy in complex vibrations in a single low-cost device comprising an array of slits. The essential morphology of the slit sensillum was examined some years ago by Barth, who made slots in sheets of Perspex or resin, then displaced them quasi-statically, and measured the changes in width at the mid-length of each slit as the sheet was deformed in-plane at a range of angles from parallel to orthogonal to the longitudinal axis of the slots (Barth *et al.*, 1984; Barth and Pickelmann, 1975). Despite this proof of principle, there has been no systematic study of the effect, for example by detailed computer simulation, nor any attempt to research its application in low-cost, distributed biomimetic sensors. Barth has recently started to model slit sensilla using finite elements.

More is known about the campaniform sensillum of insects than about the slit sensillum of spiders because it is larger and insects are much easier to experiment on. The campaniform sensillum is essentially a hole in a plate with a bell-shaped (hence 'campaniform') cap or plaque suspended in its centre. The geometry and mechanical properties of the suspension cause the cap to move up and down as the hole changes its dimensions when the plate is stretched, compressed, bent or twisted. Thus the cap system rotates deformation in the plane of the plate through 90°, allowing the deformation to be detected, by an associated sensory cell, out of the plane of the plate. The sensillum, together with its associated sensory and nerve cells, forms a simple yet sensitive mechanism which is capable of detecting displacements of the order of 1 nm (Zill and Moran, 1981) though whether this sensitivity is achieved at the mechanical or nervous level is unknown (Fayyazuddin and Dickinson, 1996). The deformations can be due to environmental loads, due to the weight of the insect's body, due to the actions of its muscles, or all three at once (Delcomyn *et al.*, 1996; Zill *et al.*, 2004). The morphology of the sensillum in insects suggests that greater sensitivity can be achieved by arranging holes in a regular pattern; that if the hole is oval it can be aligned to sense specific strain directions; and that by controlling the shape of the hole or its relationship with other holes it can have a tuned response to the rate of change in strain with time. The sensilla often occur in groups with a common orientation. Presumably such groups of sensilla are more sensitive than a single sensillum, and may also provide information regarding the direction of origin of the time-varying strain. We are currently extending the study of Skordos *et al.* to examine the design of arrays of holes in insect skeletons.

10.7   DEVELOPING AN INTERFACE BETWEEN BIOLOGY AND ENGINEERING

Although the adaptive structures which this book is addressing are confined to the realms of hi-tech, they are being developed in a forcing house which will become our everyday technology. So I find no problem in presenting an analysis which is more wide-ranging.

10.7.1   A Catalogue of Engineering

For the last five years I and a few colleagues have been grappling with a Russian system of inventive problem solving called TRIZ – *Teorija Reshenija Izobretatel'skih Zadach* – which was developed in the second half of the last century by Genrich Altshuller and Rafik Shapiro (Altshuller, 1999). TRIZ is a collection of tools and techniques that ensures accurate definition of a problem at a functional level and then provides strong indicators towards successful and often highly innovative solutions. I do not intend to describe this system, fascinating and powerful though it is. However, one of the tools characterises a problem by a pair of opposing or conflicting characteristics (typically 'What do I want?' and 'What is stopping me getting it?', but Hegel's *thesis* and *antithesis* will do as well, suggesting that it is a form of dialectic process) which can be compared with similar pairs of characteristics derived from other, solved, problems derived from the examination and analysis of more than 3 million significant patents. The characteristics are chosen from a list of 39 features, which covers most of the things that are needed or could go wrong. The pair of conflict characteristics provides coordinates to a list of 40 inventive principles (Hegelean *syntheses*) which cover, more or less, the principles which have been found at the heart of these successful (and therefore inventive) patents. This system therefore provides a compendium of technical best practice and a means of interrogation of technology.

10.7.2 Challenging Engineering with Biology

The TRIZ matrix has been developed over the past 40 years, drawing on the efforts of many people and (reputedly) about 3 million patents. In response we have analysed some 500 biological phenomena, covering over 270 functions at least three times each at different levels of hierarchy. In total we have analysed about 2500 conflicts and their resolutions in biology, sorted by levels of complexity (Vincent *et al.*, 2005). Even so, this is less than a thousandth of the data contributing to the engineering TRIZ system. To enable us to process this information we established a logical framework (Bogatyreva *et al.*, 2004) captured by the mantra: *Things do things somewhere*. This establishes six fields of operation in which all actions with any object can be executed: *Things* (substance, structure) includes hierarchically structured systems, that is the progression subsystem–system–supersystem; *do things* (requiring energy and information) implies also that energy needs to be regulated; *somewhere* (space, time). These six operational fields reorganise and condense the TRIZ classification of both the features used to generate the conflicts and the inventive principles (Table 10.3). This more

Table 10.3   TRIZ PRIZM

Operation fields that should be improved	Operation fields that cause problems					
	Substance	Structure	Time	Space	Energy/Field	Information/ Regulation
Substance	6, 10, 26, 27, 31, 40,	27	3, 27, 38	14, 15, 29, 40	10, 12, 18, 19, 31	3, 15, 22, 27, 29
Structure	15	18, 26	27, 28	1, 13	19, 36	1, 23, 24
Time	3, 38	4, 28	10, 20, 38	5, 14, 30, 34	19, 35, 36, 38	22, 24, 28, 34
Space	8, 14, 15, 29, 39, 40	1, 30	4, 14	4, 5, 7, 8, 9, 14, 17	6, 8, 15, 36, 37	1, 15, 16, 17, 30
Energy/ Field	8, 9, 18, 19, 31, 36, 37, 38	32	6, 19, 35, 36, 37	12,15, 19, 30, 36, 37, 38	14, 19, 21, 25 36, 37, 38	2, 19, 22
Information/ regulation	3, 11, 22, 25, 28, 35	30	9, 22, 25, 28, 34	1, 4, 16, 17, 39	2, 6, 19, 22, 32	2, 11, 12, 21, 22, 23, 27, 33, 34

general TRIZ matrix (which we name PRIZM – *Pravila Reshenija Izobretatel'skih Zadach Modernizirovannye* – translated as 'The Rules of Inventive Problem Solving, Modernised') is populated with the relevant inventive principles (IPs) taken from the original matrix. We add to this a new matrix – BioTRIZ – in which we place the IPs of TRIZ into a new order that more closely reflects the biological route to the resolution of conflicts (Table 10.4). We can now compare the types of solution with particular pairs of conflicts which are arrived at in technology via classical TRIZ, and in biology. Although the problems commonly are very similar, the inventive principles that nature and technologies use to solve problems can be very different.

Table 10.4   Bio-TRIZ PRIZM

Operation fields that should be improved	Operation fields that cause problems					
	Substance	Structure	Time	Space	Energy/ Field	Information/ Regulation
Substance	13, 31, 15, 17, 20, 40	1, 2, 3, 15, 24, 26	15, 19, 27, 29, 30	15, 31, 1, 5, 13	3, 6, 9, 25, 31, 35	3, 25, 26
Structure	1, 10, 15, 19	1, 15, 19 24, 34	1, 2, 4	10	1, 2, 4	1, 3, 4, 15, 19, 24, 25, 35
Time	1, 3, 15, 20, 25, 38	1, 2, 3, 4, 6, 15, 17, 19	2, 3, 11, 20, 26	1, 2, 3, 4, 7, 38	3, 9, 15, 20, 22, 25	1, 2, 3, 10, 19, 23
Space	3, 14, 15, 25	2, 3, 4, 5, 10, 15, 19	1, 19, 29	4, 5, 14, 17, 36	1, 3, 4, 15, 19	3, 15, 21, 24
Energy/ Field	1, 3, 13, 14, 17, 25, 31	1, 3, 5, 6, 25, 35, 36, 40	3, 10, 23, 25, 35	1, 3, 4, 15, 25	3, 5, 9, 22, 25, 32, 37	1, 3, 4, 15, 16, 25
Information/ regulation	1, 6, 22	1, 3, 6, 18, 22, 24, 32, 34, 40	2, 3, 9, 17, 22	3, 20, 22, 25, 33	1, 3, 6, 22, 32	3, 10, 16, 23, 25

The numbers in this table and Table 10.4 refer to the list of 40 inventive principles of TRIZ, which together represent the vast bulk of manipulations which are used in technology. The two tables therefore represent the ways in which conflicts (thesis–antithesis pairs classified according to the operating fields on the two axes) are resolved in biology and technology and can therefore form a basis for comparing problem solving in the two areas.

10.7.3 Adaptive Structures – The TRIZ Route

We wish to have an adaptive structure (*thesis*). The *antithetical* characteristics listed for the TRIZ matrix are shape, stability, reliability, complexity and convenience. These might all be compromised by the desire to incorporate adaptability. Each of these five generates a conflict pair with which to mine the TRIZ matrix. The *synthetic* IPs which occur most frequently are *parameter change* and *segmentation*. *Parameter change* is the commonest IP in the matrix, and recommends change in the physical state (e.g. to gas, liquid or solid), concentration, density, degree of flexibility, temperature, volume, pressure or any other parameter. *Segmentation* recommends dividing an object into independent parts; making it sectional or able to be dismantled; increasing the degree of fragmentation or segmentation. Both of these seem reasonable enough and, when combined, will have the effect of changing an inert and unchanging structure into an adaptive one by giving control over small or separate regions of the whole structure. We can run the same test again, this time interpreting 'adaptive' in terms of changing shape. Against this we again arrange stability, reliability, complexity and convenience. This gives another set of IPs, with *segmentation* once again at the top of the list. Other useful suggestions are to take *preliminary anti-action* (which proposes that when it is necessary to perform an action with both harmful and useful effects, this should be replaced with anti-actions to control harmful effects; pre-stress in opposition to known undesirable working stresses) which is sensible enough, and to use *dynamics*, closely allied to segmentation.

The abbreviated PRIZM matrix is necessarily more generalised. From the engineering version, taking as the *thesis* information, space and structure, and as the *antithesis* structure, we derive the IPs *segmentation, flexible shells and thin films*. Space and substance, matched against each other, most usefully yield *curvature, dynamics, composite materials*. From the BioTRIZ matrix with the same four pairs of conflict pairs we get a much richer set of recommendations, reflecting partly the multifunctionality of biology, and partly the wide array of means for achieving a particular function. The outstanding IPs are then *segmentation* (again) and *dynamics* (again) but also, quite strongly, *local quality*. This is an IP which is far more common in natural systems than in engineering (Vincent, 2005) and which is characterised as 'change an object's structure, action, environment, or external influence/impact from uniform to non-uniform; make each part of an object function in conditions most suitable for its operation; make each part of an object fulfil a different and/or complementary useful function'. This, allied with segmentation, suggests that more advantage will be gained by greater subdivision of the adaptive structure, and the allocation of a wider set of functions to each component of the structure. This would also accord with one of the other TRIZ tools showing the evolution of technology. This shows, in a number of different series, that structures become more complex and sub-specialised as they are developed; however, the idea of *local quality* suggests that more than one function will eventually be performed by each component, thus reducing the part count and increasing reliability without compromising versatility. According to the same technical evolution series, eventually the structures will become locally controlled energy fields with instant and totally adaptive control. We could start working on that now.

10.7.4 Materials and Information

When we compare the parameters which are manipulated in biology and engineering in order to solve a problem, we are presented with the shocking result that whereas in biological systems, at all sizes and degrees of hierarchy, energy is the parameter of choice in only 5 % of cases, in engineering systems it can be the parameter of choice in up to 70 % of cases. This is the peak with problems whose solution is reached by manipulations on the nanometer to micrometer scale, but even at larger scales energy figures much more prominently than in biological systems (Vincent *et al.*, 2006). Obviously this is undesirable considering the current state of the world's climate and the projected availability of energy. If energy is the most important parameter in engineering, what are the most important ones in biology, and can we move over to them instead? For if biology is to be able to give pointers to more effective strategies for survival, engineering is going to have to change, and to change in the direction of the best adapted systems for life on Earth.

Adaptive materials and structures show great promise in this area, since the main parameter which biology uses is *information*, closely followed by *structure* (Vincent *et al.*, 2006). These are the two most important parameters in adaptive structures. How should we capitalise on this, and perhaps even generalise to use adaptive materials and structures as the basis of the new engineering? We need to understand how biology manipulates these parameters to produce the desired results.

Biological systems have developed relatively few synthetic processes. Compared with engineering, which has more than 300 polymers, plus variants and blends, in biology there are only two main polymers: proteins and polysaccharides. Proteins can have incredibly large amounts of information built into them by varying the sequence of the 20 commonly found amino acids of which they are comprised, and there are many non-biogenic amino acids which can be used in abiotic proteins. These proteins form a major part of our skeleton, integrate and drive our metabolism, provide us with muscles, and so on and so forth. The main tricks of polysaccharides are a high degree of branching and the ability to bind water. So they provide cheap skeletal material and lubrication. Additionally, ribbon-like polysaccharides provide our most important fibres – cellulose and chitin. Biology then controls size by adding levels of hierarchy, possibly in a fractal manner (Prusinkiewicz and Lindenmayer, 1996; West *et al.*, 1997). A typical series would be molecule, organelle, cell, tissue, organ, organism, population, ecosystem. But there are subsystems of hierarchy; for instance, hair has six levels starting from molecular. At each level of hierarchy the functionality is controlled partly by the chemistry and partly by the structure – which is driven by the chemistry. And the chemistry is driven by information, accreted and ordered by interactions between the internal order of the organism (driven by DNA) and the external chaos – the environment. There is no such complexity in engineering where, if you want to make something bigger, all you can do is use more material in larger lumps. So biology has two further parameters which can be manipulated to be adaptive: *shape* and *hierarchy*.

We now note that engineering materials can be mapped with property dimensions such as mechanical, thermal, electrical, optical and cost. These maps show significant gaps in property space, which can sometimes be filled with hybrids of two or more materials (A, B) or of material and space ($= A + B + shape + scale$) (Ashby and Brechet, 2003). Particulate and fibrous composites are examples of one type of hybrid, but there are also sandwich structures, foams, lattice structures and others. The structural variables expand the design space of homogeneous materials, allowing the creation of new materials with specific property profiles. The next steps, of course, are to imbue these structures with adaptiveness, and then to mix-and-match in a hierarchical manner. Taking a cue from biology, the integration between levels in a hierarchical structure is probably most cheaply managed using a fractal design. Although it can be difficult and expensive to make a successful hybrid, so is the alternative of developing a new material. Both routes involve exploration of property space; the hybrid will be more likely to deliver the required properties, but the quality may be compromised by factors such as chemical incompatibility of the components. We already have some tools to short-circuit this process: for instance, a database of composites, of reinforcing fibres, chemistries and choice of structure; these methods allow promising hybrids to be identified. To this we need to add hierarchy. It is significant that some of the most efficient structures, such as airships and the Eiffel Tower, are hierarchical .(Lakes, 1993) and that the compressive strength of low-density hierarchical materials can be thousands of times greater than that of a conventional cellular material of the same density. It is clear that we should be exploring the property space of such materials in a much more systematic manner.

10.8   ENVOI

The outcome of this essay is that although there are many mechanisms in biology which could be adapted into an engineering environment, a better approach would be to adapt engineering to a set of design rules derived from biology. This does not mean that we have to invent a new form of engineering; we have to develop a new way of using the techniques we already have. Our TRIZ mapping of the way biology solves the technical problems found in engineering shows that there is only a 12 % overlap between biology and engineering, even though the underlying principles are very similar. It would seem that a significant part of the current approach to adaptive structures falls within that 12 %. As a corollary it would be interesting to compare the energy usage of an adaptive structure performing functions similar to those of a more conventional engineering structure. After all, although Darwin's hypothesis led to the aphorism 'The survival of the fittest', it also leads to 'The survival of the cheapest' (Vincent, 2002), and leaner is, as we all know, fitter.

ACKNOWLEDGEMENTS

I thank Biruta Kresling, George Jeronimidis and Olga and Nikolay Bogatyrev for their help and advice.

REFERENCES

Altshuller, G. (1999). *The innovation algorithm, TRIZ, systematic innovation and technical creativity*. Worcester, MA: Technical Innovation Center Inc.

Ashby, M.F. and Brechet, Y.J.M. (2003). Designing hybrid materials. *Acta Materialia* **51**, 5801–5821.

Barth, F.G., Ficker, E. and Federle, H.-E. (1984). Model studies on the mechanical significance of grouping in compound spider slit sensilla (Chelicerata, Araneida). *Zoomorphology* **104**, 204–215.

Barth, F.G. and Pickelmann, P. (1975). Lyriform slit sense organs modelling an arthropod mechanoreceptor. *Journal of Comparative Physiology* **103**, 39–54.

Baurecht, D. and Barth, F.G. (1992). Vibratory communication in spiders I. Representation of male courtship signals by female vibration receptor. *Journal of Comparative Physiology A - Sensory, Neural and Behavioural Physiology* **171**, 231–243.

Beukers, A. and Hinte, E.v. (1998). *Lightness: the inevitable renaissance of minimum energy structures*. Rotterdam: 010 Publisher.

Bogatyreva, O., Shillerov, A. and Bogatyrev, N. (2004). Patterns in TRIZ contradiction matrix: integrated and distributed systems. In *4th Etria Symposium*.

Darwin, C. (1875). *Insectivorous plants*. London: John Murray.

Dawson, C., Vincent, J.F.V. and Rocca, A.-M. (1997). How scales open in female pine cones. *Nature* **390**, 668.

Delarue, J.M. (1991). Minimal folding configurations. In *Proceedings of the Second International Symposium Sonderforschungsbereich 230 Part 2*, pp. 31–41.

Delcomyn, F., Nelson, M.E. and Cocatre-Zilgien, J.H. (1996). Sense organs of insect legs and the selection of sensors for agile walking robots. *International Journal of Robotics Research* **15**, 113–127.

Fagerberg, W.G. and Howe, D.G. (1996). A quantitative study of tissue dynamics in Venus's flytrap *Dionaea muscipula* (Droseraceae). II. Trap reopening. *American Journal of Botany* **83**, 836–842.

Fagerberg, W.R. and Allain, D. (1991). A quantitative study of tissue dynamics during closure in the traps of Venus's flytrap *Dionaea muscipula* Ellis. *American Journal of Botany* **78**, 647–657.

Fayyazuddin, A. and Dickinson, M.H. (1996). Haltere afferents provide direct, electrotonic input to a steering motor neuron in the blowfly, *Calliphora*. *Journal of Neuroscience* **16**, 5225–5232.

Forterre, Y., Skotheim, J.M., Dumais, J. and Mahadevan, L. (2005). How the Venus flytrap snaps. *Nature* **433**, 421–425.

Guest, S.D. and Pellegrino, S. (1992). Inextensional wrapping of flat membranes. In *Proceedings of the International Seminar in Structural Morphology*, pp. 203–215.

Haas, F. and Beutel, R.G. (2001). Wing folding and functional morphology of the wing base in Coleoptera. *Zoology* **104**, 123–141.

Herbert, R.C., Young, P.G., Smith, C.W., Wootton, R.J. and Evans, K.E. (2000). The hind wing of the desert locust (*Schistocerca gregaria* Forskål). III. A finite element analysis of a deployable structure. *Journal of Experimental Biology* **203**, 2945–2955.

Hodick, D. and Sievers, A. (1989). On the mechanism of trap closure of venus flytrap *Dionaea muscipula* Ellis. *Planta* **179**, 32–42.

Kobayashi, H., Kresling, B. and Vincent, J.F.V. (1998). The geometry of unfolding tree leaves. *Proceedings of the Royal Society B* **265**, 147–154.

Kresling, B. (1991). Folded structures in nature - lesson in design. In *Proceedings of the 2nd International Symposium SFB 230 Part 2*, pp. 155–161.

Kresling, B. (2000). Coupled mechanisms in biological deployable structures. In *Deployable structures: theory and applications*, ed. S. Pellegrino and S.D. Guest, pp. 229–237. London: Kluwer Academic.

Lakes, R.S. (1993). Materials with structural hierarchy. *Nature* **361**, 511–515.

Miura, K. (1993). Concepts of deployable space structures. *International Journal of Space Structures* **8**, 3–16.

Probine, M.C. and Preston, R.D. (1962). Cell growth and the structure and mechanical properties of the wall in internodal cells of *Nitella opaca*. II Mechanical properties of the walls. *Journal of Experimental Botany* **13**, 111–127.

Prusinkiewicz, P. and Lindenmayer, A. (1996). *The Algorithmic Beauty of Plants*. Berlin: Springer.

Skordos, A., Chan, C., Jeronimidis, G. and Vincent, J.F.V. (2002). A novel strain sensor based on the campaniform sensillum of insects. *Philosophical Transactions of the Royal Society A* **360**, 239–254.

Smith, C.W., Herbert, R., Wootton, R.J. and Evans, K.E. (2000). The hind wing of the desert locust (*Schistocerca gregaria* Forskål). II. Mechanical properties and functioning of the membrane. *Journal of Experimental Biology* **203**, 2933–2943.

Stuhlman, O. (1948). A physical analysis of the opening and closing movements of the lobes of Venus's flytrap. *Bulletin of the Torrey Botanical Club* **75**, 22–44.

Thom, R. (1975). *Structural stability and morphogenesis: an outline of a general theory of models*. Reading, MA: WA Benjamin.

Vincent, J.F.V. (2002). Survival of the cheapest. *Materials Today* 28–41.

Vincent, J.F.V. (2005). Deconstructing the design of a biological material. *Journal of Theoretical Biology* **236**, 73–78.

Vincent, J.F.V., Bogatyreva, O., Pahl, A.-K., Bogatyrev, N. and Bowyer, A. (2005). Putting biology into TRIZ: a database of biological effects. *Creativity and Innovation Management* **14**, 66–72.

Vincent, J.F.V., Bogatyreva, O.A., Bogatyrev, N.R., Bowyer, A. and Pahl, A.-K. (2006). Biomimetics—its practice and theory. *Journal of the Royal Society Interface* **3**, 471–482.

Vincent, J.F.V. and Jeronimidis, G. (1992). Mechanical design of fossil plants. In *Biomechanics in Palaeontology*, ed. J.M.V. Rayner and R.M. Wootton, pp. 21–36. Cambridge: The University Press.

von Guttenberg, H. (1959). Die physiologische Anatomie seismonastisch reaktionsfähiger Organe. In *Handbuch der Pflantzenphysiologie*, ed. W. Ruhland, pp. 175–191. Berlin: Springer.

von Guttenberg, H. (1971). Bewegungsgewebe und Perzeptionsorgane. In *Handbuch der Plantzenanatomie*, ed. W. Zimmermann, pp. 171–175. Berlin: Verlag Gebr. Bornträger.

West, G.B., Brown, J.H. and Enquist, B.J. (1997). A general model for the origin of allometric scaling laws in biology. *Science* **276**, 122–126.

Williams, S.E. and Bennett, A.B. (1982). Leaf closure in the Venus Flytrap: an acid growth response. *Science* **218**, 1120–1122.

Wootton, R.J., Evans, K.E., Herbert, R. and Smith, C.W. (2000). The hind wing of the desert locust (*Schistocerca gregaria* Forskål). I. Functional morphology and mode of operation. *Journal of Experimental Biology* **203**, 2921–2931.

Zill, S., Schmitz, J. and Buschges, A. (2004). Load sensing and control of posture and locomotion. *Arthropod Structure & Development* **33**, 273–286.

Zill, S.N. and Moran, D.T. (1981). The exoskeleton and insect proprioception I. Responses of tibial campaniform sensilla to external and muscle-generated forces in the American cockroach, *Periplaneta americana*. *Journal of Experimental Biology* **91**, 1–24.

Index